Phenomenology and Psychological Science

HISTORY AND PHILOSOPHY OF PSYCHOLOGY

Series Editor: **Man Cheung Chung**, *University of Plymouth, Plymouth, United Kingdom*

DESTINED FOR DISTINGUISHED OBLIVON
The Scientific Vision of William Charles Wells (1757–1817)
Nicholas J. Wade

PHENOMENOLOGY AND PSYCHOLOGICAL SCIENCE
Historical and Philosophical Perspectives
Edited by Peter D. Ashworth and Man Cheung Chung

REDISCOVERING THE HISTORY OF PSYCHOLOGY
Essays Inspired by the Work of Kurt Danziger
Edited by Adrian Brock, Johann Louw, and Willem van Hoorn

UNFOLDING SOCIAL CONSTRUCTIONISM
Fiona J. Hibberd

A Continuation Order Plan is available for this series. A continuation order will bring delivery of each new volume immediately upon publication. Volumes are billed only upon actual shipment. For further information please contact the publisher.

Phenomenology and Psychological Science
Historical and Philosophical Perspectives

Edited by

Peter D. Ashworth
Sheffield Hallam University
Sheffield, United Kingdom

and

Man Cheung Chung
University of Plymouth
Plymouth, United Kingdom

Springer

Peter D. Ashworth
Sheffield Hallam University
Deepcar S36 2NR
Sheffield, United Kingdom
e-mail: p.d.ashworth@shu.ac.uk

Man Cheung Chung
Clinical Psychology Teaching Unit
University of Plymouth
Plymouth PL4 6AA
United Kingdom
e-mail: m.chung@plymouth.ac.uk

ISBN 978-1-4419-2219-9 e-ISBN 978-0-387-33762-3

Printed in the United States of America. (TB/MVH)

9 8 7 6 5 4 3 2 1

springer.com

CONTRIBUTORS

Peter D. Ashworth is Professor of Educational Research at Sheffield Hallam University, Sheffield, UK

Man Cheung Chung is Reader in Psychology at the University of Plymouth, Plymouth, UK

Karin Dahlberg is Professor of Nursing Studies at the University of Växjö, Växjö, Sweden

Amedeo Giorgi was founding Editor of the *Journal of Phenomenological Psychology* and is Professor of Psychology at Saybrook Graduate School, San Francisco, California, USA

Barbro Giorgi is Professor and Director of MA Research at JFK University, San Francisco, California, USA

Miles Groth is Professor of Psychology at Wagner College, Staten Island, New York, USA

Stuart Hanscomb is Lecturer in Philosophy at the Crichton Campus of Glasgow University, Dumfries, UK

Alec Jenner is Emeritus Professor of Psychiatry at the University of Sheffield, Sheffield, UK

Paul S. MacDonald is Senior Lecturer in Philosophy at Murdoch University, Perth, Australia

CONTENTS

1. The Meeting Between Phenomenology and Psychology 1
Man Cheung Chung and Peter D. Ashworth

Initial Contact Between Phenomenology and Psychology 2
The Question of Phenomenological Psychological Method 5
Husserl's Transcendental Subjectivity 5
Heidegger's Critique and The Existentialist Turn 6
Boss and Daseinsanalysis and Contemporary Existentialists in Psychology 8
A Long Story 9
References 9

2. Introduction to the Place of Phenomenological Thinking in the History of Psychology 11
Peter D. Ashworth

The early Experimental Psychology of Experience 12
Early Twentieth Century Theoretical Ferment and the Emergence of Behaviorism 14
Husserl's Phenomenology 17
The Existentialist Critique of Husserl's Phenomenology 24
The Legacy of Husserl for Psychology 28
The Early Confluence with Gestalt Psychology 29
American "Psychological Phenomenology" 32
Contemporary Phenomenological Psychology 36
Conclusion 41

Note 42
References 42

3. The Value of Phenomenology for Psychology 45
Amedeo Giorgi

Introduction 45
The Phenomenological Approach to Consciousness 46
Psychology's Approach to Consciousness in the Founding Years . . . 54
The Approach to Consciousness of the Functional School 60
James's Approach to the Phenomenon of Consciousness 61
The Approach to Consciousness on the Part of Act Psychology 63
Conclusion 66
Note 67
References 67

4. Can an Empirical Psychology Be Drawn from Husserl's Phenomenology? 69
Barbro Giorgi

Introduction 69
The Specific Procedures of the Method 71
How Phenomenological Concepts Guide the Research 73
Conclusion 87
References 87

5. Did Husserl Change His Mind? 89
Karin Dahlberg

Introduction 89
Phenomenology as "Going Back to Things Themselves" 90
Transcendental Phenomenology 91
Intentionality 93
Self Reflection and Self Awareness 94
Conclusions 96
Note 100
References 100

6. Husserl against Heidegger against Husserl 101
Paul S. MacDonald

Introduction 101
Five Foci of Critique 110

Summary 120
References 121

7. The Influence of Heidegger on Sartre's Existential Psychoanalysis 123
Miles Groth

Preliminaries 123
Sartre's Existential Psychoanalysis 124
Heidegger's Influence on Sartre's Existential Psychoanalysis 130
Heidegger and Sartre's Existential Psychoanalysis 134
Epilogue: Change 136
Notes 137
References 143

8. Medard Boss' Phenomenologically Based Psychopathology 147
F. A. Jenner

Personal Background 147
Historical Setting and Training of Medard Boss 149
Presumptions 154
Sigmund Freud's Impact on Medard Boss 155
Summary 167
References 167

9. Contemporary Existentialist Tendencies in Psychology 169
Stuart Hanscomb

How Existentialism and Psychology Meet 169
Methodology 188
Notes 191
References 193

10. Conclusion: Phenomenology and Psychological Science . . . 197
Peter D. Ashworth and Man Cheung Chung

References 205

Name Index 207

Subject Index 213

CHAPTER 1

THE MEETING BETWEEN PHENOMENOLOGY AND PSYCHOLOGY

MAN CHEUNG CHUNG and PETER D. ASHWORTH

> Scientific points of view, according to which my existence is a moment of the world's, are always both naïve and at the same time dishonest, because they take for granted, without explicitly mentioning it, the other point of view, namely that of consciousness, through which from the outset a world forms itself round me and begins to exist for me.
>
> (Merleau-Ponty, 1945/1962, p. ix)

Merleau-Ponty's statement applies exactly to the divergence between psychological science as it currently exists and any phenomenologically based human study. Even in circles in which it has become the vogue to employ "qualitative methods" there is often an underlying scientism of the kind Merleau-Ponty indicated. Human experience, or discursive action, is seen as part of a causal nexus, a set of variables within the "world." The chapters in this volume explore the meaning of the "other point of view, that of consciousness."

Some chapters focus on the history of psychology and the ways in which various psychologists (often rather isolated voices) developed ways of researching and theorizing that took account of at least some aspects of the phenomenological tradition in philosophy. Other chapters explore key philosophical debates within phenomenology itself—though always with an emphasis on their meaning for the realm of psychology. But in addition to the historical and technically philosophical themes, we include chapters that (though referring in some depth to the arguments within the phenomenological tradition of thought) indicate how phenomenologically sound work can be carried out in the realm of psychology.

Understanding human "nature," some may argue, is a futile exercise because the complexity of human beings removes them from scholarly comprehension (though somehow we manage to conduct our practical daily relationships more-or-less adequately). While many psychologists, including the editors and the other authors in the present volume, would not dispute the fact that human beings are extremely complex, we nevertheless still hold a belief that some aspects of human action are open to fruitful exploration and indeed can be understood. What drives us to hold rather than relinquish such a belief is perhaps our curiosity about who we are and why we behave the way we do. Moreover, we are driven by our conviction that there are in fact rigorous approaches that can help us investigate and understand some aspects of being human. Whether seen as complimentary to other approaches (the "scientific" ones which to which Merleau-Ponty pointed, among them) or whether seen as uniquely qualified as a methodology for human studies, one such approach is that informed by phenomenology. The overall aim of this book is to articulate the extent to which phenomenology can assist us in understanding some aspects of human psychology.

The careful reader will have already noted the areas in which debate is likely to happen. Do we call phenomenological approaches "scientific," or do we reserve this term—together with its high contemporary status—for the explanation of human action in terms of variables externally assessable? Is the study of experience a line of research which is going to lead to understanding or explanation or both? Is research aimed at the evocation of "inner" experience likely to be lost in individual subjectivity? And what is the relation between experience and discourse? Some of these questions will resonate in the chapters that follow.

INITIAL CONTACT BETWEEN PHENOMENOLOGY AND PSYCHOLOGY

The book begins by setting the scene for the initial contact between phenomenological philosophy and psychological science. Ashworth (see Chapter 2) draws our attention to the fact that mainstream psychology, at the beginning, was concerned with the study of experience, and this should have meant that the ground was well prepared for the arrival of phenomenology; psychology was engaged with topics which would relate well to the approach of phenomenologists. Despite this, Ashworth tells us that the behaviorist revolution drove a wedge between most psychological sciences and the emerging phenomenological philosophy of Husserl (1913/1983).

On the face of it, the detailed study of *the experience of being conscious of something* seems to be a kind of psychology. But Husserl had a philosophical purpose in founding phenomenology, not an immediately psychological one. Ashworth shows that Husserl's work, like that of the later disputatious members of his school, needs to be worked through carefully if its relevance to psychology is to be fully grasped. The early aim of Husserl for his phenomenology is apparently that it would be a reflective philosophical discipline that would provide a conceptual underpinning for the different sciences and scholarly disciplines. This would include psychology. But Ashworth traces the dissent of Heidegger and later existential phenomenologists from this approach. Heidegger rejected, in particular, the Husserlian assumption that the philosopher could be sufficiently detached from the everyday world to be able to describe each phenomenon in its purity. Instead he argued for—and practiced—an existential and hermeneutic phenomenology. We learn from Ashworth and Chung in Chapter 10, however (where this dispute is taken up again), that Husserl and Heidegger should both be seen as *transcendental philosophers*—both locate their analyses in the world of conscious experience. This issue of phenomenological method is absolutely pervasive throughout the book, and we shall later find that it dominates the discussion of Chapters 5, 6 and 7, in particular.

Heidegger proposed "the world" as a central concept. Though there is a sense in which this establishes the primacy of "my" perspective, nevertheless, it can be shown to have a definite structure which all of us "in" the world necessarily share. Elsewhere in the book, authors use the cognate term "lifeworld."

Having laid out some of the developments in phenomenological philosophy, Ashworth discusses the relationship between phenomenology and some developments in psychology, notably Gestalt psychology (especially the work on perception), and some related approaches in social psychology due to Heider and to Kurt Lewin. In the United States, a stream of work related to phenomenology emerged after the Second World War. The idiographic personality theory of Gordon Allport has affinities with existential phenomenology. To Allport, the whole person is a unique entity and needs to be understood as a coherent whole. He or she is not simply a collection of parts or elements. But Ashworth sees a distinction between Allport and the phenomenologists due to Allport's lack of any insistence on the primacy of the individual's own perspective. More committed to the phenomenological approach are Robert Macleod and Snygg and Combs, who show considerable awareness of Husserl and the epochē—the determination to turn attention to conscious experience and the bracketing of the question of its relation to reality.

American "humanistic psychologists" were also in line broadly with phenomenological thinking, especially in its existential version. These psychologists were concerned with issues pertaining to authenticity and freedom, despite the fact that they, on the whole, tended not to engage in the methodological rigor of phenomenology.

Ashworth concludes his chapter by introducing us to some of the contemporary voices within psychology which urge phenomenological methodology on the discipline. One of these is Amedeo Giorgi, the author of Chapter 3, and the founding editor of the flagship journal in this area, the *Journal of Phenomenological Psychology.*

A. Giorgi begins Chapter 3 by describing Husserl's conceptualization of consciousness. For instance, Husserl believed that consciousness is a type of being which is different from a physical "real thing"; consciousness is not given to us via sensory experience nor as a spatio-temporal entity regulated by causal laws. Consciousness can present to us, in a direct way, objects other than empirical ones. Husserl contends that we can access our consciousness through the phenomenological attitude of bracketing and reduction. The main function of consciousness, A. Giorgi tells us, is intuition and not experiencing; consciousness can know itself and such awareness is not through appearances. Consciousness is *intentional.*

A. Giorgi then turns attention to the way in which early psychologists treated consciousness. He develops the brief account by Ashworth in Chapter 2 considerably, by describing the structural psychology of which Wundt was an advocate. Despite Wundt's empirical and natural science approach, he distinguished psychology from natural sciences in that, while the former is concerned with the objects of experience dependent on experiencing subjects, the latter is concerned with the objects of experience independent of experiencing subjects. Wundt believed that inner experience and outer experience did not refer to different kinds of objects, but different ways of looking at identical objects experienced.

A. Giorgi explains that Titchener, an American student of Wundt (whose version of structural psychology was taken in the United States as authoritative), viewed psychology as a natural science with a method based on experimentation. Titchener believed that the mind does not *have* thoughts and feelings. Instead, the mind *is* thoughts and feelings. To Titchener, consciousness means the mind now, the mind of the present moment or the mind at every "now." Every consciousness is composed of a number of concrete processes (wishes, feelings, ideas, etc.) which, in turn, are composed of a number of really simple processes that are coming together. That is, Titchener took "consciousness now" to be a totality which can subsequently be broken into different basic elements. Conscious activities are processes within this totality but these activities are not related to the outside world. That is, intentionality does not play a role in Titchener's approach to psychology, which is often called a psychology of content rather than a functional psychology or a psychology of act.

The functional school's approach to the study of consciousness was different again. Functionalism was essentially naturalistic. Anti-dualistic, consciousness was simply seen to be a matter of evolutionary development. Nevertheless, the functionalist's approach does entail some of the features of psychological research that phenomenology values, in that, for example, consciousness, according to the functional psychologists, should be understood within the context of person–environment relationships.

Also moving away from the dualistic approach, William James talked of "radical empiricism." To him, there is only one type of "stuff" (pure experience) of which the world is made. His approach to consciousness is effectively phenomenological *avant la lettre*. He described the stream of consciousness as personal, selective, constantly changing, continuous and related to independent objects. Through introspection, consciousness becomes aware of itself.

THE QUESTION OF PHENOMENOLOGICAL PSYCHOLOGICAL METHOD

The chapters so far have revealed a direction that the influence of Husserlian thought takes with respect to psychology. But the immediate question of research psychologists—who are hugely practical people and will not entertain a theoretical stance, however convincing, without a clear approach for empirical research—will certainly be *But how is phenomenological psychology carried out?* Barbro Giorgi, in chapter 4, describes a method of research based firmly on phenomenology and which can be applied to empirical work in psychology. This method aims to discover, articulate, and make explicit the participant's lived psychological meanings. The practical steps of the phenomenological psychological method are given in sufficient detail in this chapter to enable readers to make some attempt at such research themselves. (It has to be said that phenomenology will never lay out mechanical techniques of research after the style of an experimental design: the target of the research is the elucidating of experience rather than the testing of causal hypotheses.) Behind the practical steps which she describes, the classic phenomenological descriptions of the essential features of lived experience encapsulated in notions such as the lifeworld, intersubjectivity, intentionality, intuiting, pre-reflectiveness and the epochē act as guiding concepts. B. Giorgi devotes considerable attention to clarifying the relation of these to the research method.

HUSSERL'S TRANSCENDENTAL SUBJECTIVITY

While the foregoing chapters have demonstrated the relevance of Husserlian thought for psychology, his approach is not free from controversy. What follows is a chapter (Chapter 5) which aims to address one of these controversies, namely, Husserl's view on transcendental subjectivity. Dahlberg here wishes to explore two broad questions: Did Husserl change his mind on his view on transcendental subjectivity? Did his followers take up the idea of transcendence at all? She argues that the answer to question one is "no" and to question two is "yes."

Did Husserl change his mind? According to Dahlberg, through a focus on the transcendental, Husserl intended to go beyond the natural attitude (the uncritical and unanalytical attitude with which we go about our practical activities in the world), in order to understand the world. However, there are limitations and constraints in his transcendental subjectivity because it is inevitably approached by real persons who are immersed in the world of experience and cannot fully detach from it. That is, it is impossible to bracket off all of our pre-understanding or pre-assumptions so as to turn back with a clear view on the lifeworld (we can never be completely free of our own prejudice). In turn, this means that Husserl's ultimate wish to discover pure consciousness is unachievable and unrealistic.

Although it is impossible to obtain the purity of transcendental subjectivity, as Dahlberg remarks, "the path between natural attitude and the point of pure transcendence is accessible to us and provides an entry point from which we can develop an approach of critical scrutiny for research." When Husserl spoke about transcendental subjectivity, he not only had this notion of pure consciousness in mind, but also the notions of self-reflection and self-awareness (i.e., our human ability to think about and reflect upon our own consciousness). Thus, while the human lifeworld is characterized by a natural attitude, within the lifeworld, a critical attitude is attainable that involves the processes of reflecting on that of which we are aware. Through reflection, consciousness, which is directed towards the world, turns towards the self. Consequently, consciousness establishes a distance between itself and the world and between itself and the natural attitude. It is possible, then, to focus more critically on the phenomenon that we are investigating.

Did his followers take up the idea of transcendentality at all? Dahlberg argues that Merleau-Ponty and Gadamer were clearly thinking in line with Husserl. For example, they believed that philosophers should suspend the affirmations implied in the facts given to us in our lives. However, to suspend them does not mean to deny them. Neither does this mean to deny the link which binds us to the physical, social, and cultural world in which we live. Quite the contrary, we should see and become conscious of the link through "phenomenological reduction."

HEIDEGGER'S CRITIQUE AND THE EXISTENTIALIST TURN

The criticism of Husserl which focuses on the fact that unprejudiced reflection on "pre-reflective" experience is not fully possible is especially associated with Heidegger (1927/1962, 1927/1988). Several of these lines of criticism have had the valuable effect of bringing empirical psychology and phenomenology into greater contact. So the importance of these critiques must not be underestimated because they have turned out to be vital in strengthening the relationship between phenomenological philosophy and the psychological sciences, and in nurturing the development of phenomenological psychology. The book proceedes to examine some of these critiques.

MacDonald (see Chapter 6) argues that Heidegger, Jaspers, Sartre, and Merleau-Ponty acknowledged the fact that one of their points of rupture from Husserl had to do with phenomenological method. MacDonald focuses on five principal criticisms put forward by Heidegger:

1. Over-theoretization (theoreticism): This is concerned with the idea that all forms of human attitude toward the world are construed as analogues to a theoretical attitude. One attitude is the genetically primitive. The other is the "derivative" attitude of natural science.
2. Over-intellectualization (intellectualism): This is concerned with the idea that all forms of human behavior towards the world are construed as varieties of an intentional directedness modeled on an intellectual encounter with objective aspects of mere things. This is also concerned with the fact

that the corporeal, affective and evaluative dimensions of us are derived from some basic stratum of intellectual apprehension.

3. Splitting of the ego: This is concerned with the idea that transcendental ego is separate from the empirical or mundane ego. Our consciousness is composed of two separate realms, one being anonymous, lifeless and neutral and the other being personalized, full-of-life and interested.
4. Consciousness and world separated by an abyss: The apparent gulf between the being of the conscious and that of the non-conscious (what Sartre refers to as *être pour soi* and *être en soi* requires detailed phenomenological attention.
5. Neglect of the understanding of the meaning of intentionality: Macdonald tells us that the question of the distinction between the being of the intentional act and the being of an intentional agent is much neglected.

It is hardly controversial to say that, in Husserl and Heidegger, we have the founder of phenomenology in its contemporary guise, and the instigator of the main lines of deviation. We would also say (though this *is* controversial), these lines of deviation are extraordinarily creative and productive. This positive judgment of (at least the early Heidegger) has possibly been borne out in such psychotherapeutic applications as those of the humanistic psychologists (mentioned in Chapter 2) and of Medard Boss (Chapter 8). So at this stage in the book we turn to some important phenomenologists/existentialists, who were influenced to a significant degree by Husserl and Heidegger.

Focusing on Sartre and Heidegger, Groth (Chapter 7) aims to address two questions: What was Sartre's contribution to psychology and to what extent was Sartre's psychology influenced by Heidegger's thought?

To address the first question, Groth speaks of existential psychoanalysis, which begins with the idea that human reality is a unity rather than a collection of functions. However this may be, we are never able to see or know ourselves as a unity because we are forever changing (i.e., forever "condemned" to be a choice of being). That is why existential analysis denies the notion of the unconscious and the unmediated influence of the environment upon us. Groth also talks about Sartre's notion of "bad faith" which can only be understood in terms of the fact that "the being of consciousness is the consciousness of being." According to Sartre (1943/1958), Groth argues, the orthodox psychoanalytic model of the censor which operates in our psychic apparatus is in bad faith. Groth tells us that the significant differences between existential psychoanalysis and orthodox psychoanalysis lie in their emphasis on the present versus the past, freedom versus determinism, and interminable versus terminable periods of therapy.

To address the second question, Groth points out that however influenced by Heidegger Sartre might have been, Heidegger did not think that Sartre was entirely correct in understanding some of his findings or opinions. One of Heidegger's lines of critique of Sartre revolves around the fact that Sartre read *Being and Time* as if it were a work of metaphysics which, in fact, Heidegger wanted to reject and deconstruct. Heidegger rejected existentialism partly because he thought it to be characterized by metaphysical thinking in which human beings are the focus. On the contrary, Heidegger was interested in the notion of be(ing). Groth concludes that "although Sartre freely adopts some of the terminology of *Being and Time* in *Being and Nothingness*, we must conclude that Sartre's Heidegger is not one Heidegger himself would recognize." Another criticism by Heidegger questioned whether Sartre's way of thinking can be constructed as a kind of humanism (to which Heidegger was averse and which he did not recognize in his own thought. Heidegger also criticized Sartre for failing to "recognize the essentiality of what is historical about be(ing)." Heidegger, then, saw the *Dasein* (the being of the human kind) as immersed in historical and cultural forms. Sartre—and we have to say, for the most part, Husserl—did not.

BOSS AND DASEINSANALYSIS AND CONTEMPORARY EXISTENTIALISTS IN PSYCHOLOGY

Heidegger's version of phenomenology (and—it is worth saying—we do regard the Heidegger of the period of *Being and Time* as a phenomenologist) strongly influenced the development of Daseinsanalysis, pioneered by Medard Boss. Jenner (see Chapter 8) provides a biographical sketch of Medard Boss who was closely associated with Freud, Jung, Binswanger, and Bleuler as well as Heidegger. Some of the philosophical ideas underlying Daseinsanalysis, for which Boss was indebted to Heidegger, are described. Boss disagreed with Freud in several ways. He showed, for example, that Freud's lingering faith in the ontology of 19th century physics and the natural sciences was unnecessary in the arena of psychotherapy. He also showed that Freud was mistaken in believing that narcissistic neurosis is not amendable to psychological treatment. Despite Boss's critical view of Freud, he admired Freud's technique of free association and in fact used it to help his patients to discover their own potential. According to Jenner, Boss's therapeutic approach was humane, concerned, and impressive. The underlying philosophy was also interesting, though not unproblematic. Boss had indeed helped to nurture a view of psychiatry beyond simplistic medical axioms.

Turning to more contemporary existentialists in psychology and psychotherapy, Hanscomb (Chapter 9) describes Yalom's existential notion of "ultimate concerns" (i.e., death, freedom, responsibility, willing, isolation, and meaninglessness). These concerns are not independent of each other but are interwoven, being fundamentally based upon conscious human existence. Van Deurzen-Smith's existential analysis contrasts with Yalom's in speaking of "existential dimensions": physical, social, psychological, and spiritual. Hanscomb aims to map the self with reference to these "concerns" and "dimensions." He argues that "the experience that makes best sense of all these concerns is separation (alienation), uncanniness or a sense of not-at-homeness." If it is the case that some form of alienation, uncanniness or not-at-homeness is inevitably a part of our human condition, it is not surprising that some form of anxiety is also part of this condition. Hanscomb develops his thesis in terms of our relationship with other people, the notions of freedom, guilt, and death, the notion of authenticity, and the notion of meaning.

A LONG STORY

In the final chapter, Ashworth and Chung attempt to bring, if not closure, at least some sense of summation to the book. For the story of phenomenology and psychological science is complex if not a short story, or a straightforward one. Neither let us pretend that the present volume tells the whole story. Far from it, this book provides only a taste of a very wide discussion, which has a century-long history and continues in a renewed way with the rise of post-modernism. It is the view of the contributors to this volume that the approach inspired by phenomenology is not merely historical, nor is it confined to philosophy. It has a contemporary application to psychology and one which, we believe, will be of growing importance as psychologists become more aware of that "other point of view, namely that of consciousness, through which from the outset a world forms itself round me and begins to exist for me." The fundamental theme of phenomenology in psychology is that we seize again the meaningfulness of our own lived experience.

REFERENCES

Heidegger, M. (1988). *The basic problems of phenomenology* (A. Hofstadter, Trans.). Oxford: Blackwell. (1975 posthumous version; Original lectures published 1927)

Heidegger, M. (1962). *Being and time* (J. Macquarrie & E. Robinson, Trans.). Oxford: Blackwell. (Original work published 1927)

Husserl, E. (1983). *Ideas pertaining to a pure phenomenology and to a phenomenological philosophy* First Book (F. Kersten, Trans.). Dordrecht: Kluwer. (Original work published 1913)

Merleau-Ponty, M. (1962). *Phenomenology of perception* (C. Smith, Trans.). London: Routledge and Kegan Paul. (Original work published 1945)

Sartre, J. P. (1958). *Being and nothingness* (H. E. Barnes, Trans.). New York: Philosophical Library. (Original work published 1943)

CHAPTER 2

INTRODUCTION TO THE PLACE OF PHENOMENOLOGICAL THINKING IN THE HISTORY OF PSYCHOLOGY

PETER D. ASHWORTH

The account in this chapter is intended to provide an introductory framework, allowing some of the detailed arguments of later chapters to be contextualized. The following, therefore, attempts to locate the points at which there has been contact between phenomenological philosophy and psychological science—and the various ways in which phenomenologists have argued that there *should have* been contact.

Even the most superficial reading of phenomenology would alert one to the concern that this school of thought has with experience (though the exact meaning to be given to this word no doubt requires specification). And, on the face of it, the detailed study of *the experience of being conscious of something* seems to be a kind of psychology. Nevertheless, it has to be said that, most unfortunately, phenomenological thinking has been marginal in the history of psychology. Two reasons for this need to be mentioned at the outset. Firstly, Husserl had a philosophical purpose in founding phenomenology, which was not by any means immediately psychological. His work, like that of the later disputatious members of his school, needs to be worked through carefully if its relevance to psychology is to be fully grasped. Secondly, Husserl's line of thinking emerged at a point in the history of psychology when discussion of experience as such was especially unwelcome. It is true that, when experimental psychology was founded in the second half of the nineteenth century, it was *defined* as the science of experience, nevertheless by the early years of the twentieth century, dilemmas regarding the scientific meaning of conscious experience had led to a widespread move away from this concern.

The philosophers and physiologists (in the main) who began to establish psychology as a discipline had seen the immensely impressive strides in understanding the nature of the external world made by the physical sciences. Psychology would complement this by developing a scientific understanding of the inner world of experience; this inner realm would be approached experimentally and quantitatively. We shall see what kind of research this involved, and the *behaviorist* reaction which it evoked.

THE EARLY EXPERIMENTAL PSYCHOLOGY OF EXPERIENCE

A major interest of those early experimentalists, in fact, was in discovering what precisely the relationship was between the "outer" and the "inner" worlds. (Yes, unfortunately there was an assumption that this distinction could be assumed for all practical purposes.) Gustav Fechner (1801–1887) who was, with certain reservations, regarded by the premier historian of experimental psychology, Edwin Boring (1950), as the founder of the discipline, aimed to discover the laws relating the physical nature of an external stimulus to the internal experience of the sensation it produced. Fechner's *Elemente der Psychophysik* (1860/1966) could indeed be regarded as the founding publication of experimental psychology. In it, Fechner reported his findings on such matters as the relationship between a change in light intensity and the subjective sensation of brightness. But what was the meaning of "experience" in experimental work such as Fechner's? It was limited in the extreme, and boiled down to the individual report of some aspect of a sensation. The fact that the experience of variations in brightness was within a very specific, controlled context, with a particular social meaning (and so on) was, it appears, of no interest to Fechner.

Right at the start there was scientific controversy surrounding Fechner's book. Some of it was aimed at the details of the methodology. But William James was one distinguished psychologist who regarded the whole enterprise of "psychophysics" as completely without value. However, for the most part, the human capacity to report verbally on sensations of the elementary kind investigated by Fechner ("Which light is brighter?" "The one on the left.") could, it seems, appear unproblematic given the restricted focus of interest of the experimental investigation. Later investigators developed psychological studies which had more complex aims, however. Thus, Wundt's *Physiologische Psychologie* (1874/1904) was concerned with immediate experience in terms of its discriminable elements and the manner of their inter-relationships. Wundt believed immediate experience to be made up of elements (sensations, images, and feelings) which are combined in various ways. The laboratory investigation of the nature of the elements and the laws of their inter-relationships, while systematic in the extreme and controlled at the level of stimuli, nevertheless depended on the research participants' verbal reports of their (a question-begging term) *introspections*.

Wundt's work on the "structure" of immediate experience did not by any means remain unchallenged. In particular, Brentano (1874/1995) developed a quite different approach to immediate experience, regarding it as a process or *act*, so that different kinds of experience are to be distinguished, not by the way in which they are structured in consciousness, but by the particular way in which consciousness relates to the object of experience. Judgment and perception, for instance, involve different orientations to the object. The definitive feature of conscious activity, for Brentano (and this was taken up by Husserl and the phenomenologists), was its *intentionality*, a technical term pointing to the intrinsic "relatedness" of consciousness to the object of its attention. The fact that consciousness—unlike any other process—has this attribute of intentionality was definitive. "All consciousness is consciousness of something." And psychology had the task of delineating the various ways in which consciousness could relate to its objects.

Brentano's act psychology did not gain a significant hearing outside Germany, though it has an impact on Gestalt theory. And Wundt's structural psychology with its introspectionist technique and focus on mental content, gave way to functionalism, especially in its behaviorist form in the Anglo-American world. But in the meantime the psychological descriptions of William James are of great importance.

William James and the Internal Stream of Consciousness

In volume one of James' *Principles of Psychology* (1890/1950), we have a basic psychology of experience, primarily in terms of the stream of conscious but also through the description of two meanings of "self." The thing which distinguished James's description of experience from those of Fechner and Wundt was that, whereas they were concerned to find the elements that combined together in various ways to make the totality of experience at a particular time, James rejected this atomism in favor of the attempt to describe key features of the field of awareness taken in its entirety. James described consciousness as an ongoing process, having its own themes within which the current foci of attention get their meaning. So the content of consciousness is, at a particular moment, a phase of a personal "stream." The significance of a particular object of consciousness is not just due to its reference to the external thing but is also due to its relationship to the ongoing themes of my awareness—its personal relevance to me.

James builds up a general case for the importance of what he calls the "fringe" of the focal object of our conscious experience. An object of awareness gains its meaning in large measure from the "halo of relations" with which it is connected—its "psychic overtone." Husserl later also pointed to a similar idea: the "horizon" of a phenomenon. That is, an object of awareness is affected intrinsically by the whole web of its meaningful connections within the world of experience. Choice is also a feature of consciousness for William James. Of the available objects of attention, one becomes focal at a particular time and others are reduced to the periphery of attention. Here, we have something akin to the Gestalt psychologists' distinction between the figure and ground of awareness.

James's approach to consciousness is continued in the subsequent chapter of the *Principles*, which is devoted to the self. James regards this as a very difficult topic, but he discusses in detail the distinction between the self as an object of thought (the self-concept, let us say), and the self as that *who* is aware of that self-concept. So the self is a "duplex" (as James puts it) involving both (a) the self which we can conceptualize, the self as known, the *me*, and also (b) the self as that which "has" that knowledge, the *I*. In addition, the me is shown to have a complex structure itself. So James provides a basic phenomenology of the self, which was developed by such later authors as G.H. Mead and Gordon Allport.

The basic description of awareness and self was a valuable advance. James, much later, continued the descriptive tendency of his work in a way which also employed a form of qualitative research. This was in the groundbreaking *Varieties of Religious Experience* (1902). In this book, James draws on a wide range of texts and personal accounts, which are—in an important methodological move akin in some ways to the phenomenological process of "bracketing" reality—interpreted as matters of subjective conceptualization, rather than in terms of any external reality to which the perception or conception is supposed to refer.

EARLY TWENTIETH CENTURY THEORETICAL FERMENT AND THE EMERGENCE OF BEHAVIORISM

Unfortunately, the very fruitful forms of literary-qualitative research shown in James's psychology did not remain part of mainstream psychology but were submerged in the general disillusion with Wundtian introspection. Critique of introspection took several forms and each form, it seems, gave birth to a distinct school of psychological research, but the dominant one, especially in American academic psychology, was behaviorism. This line of thinking was especially inimical to any phenomenologically-oriented approach and so it is important to note its characteristics.

Historically, then, behaviorism began as a methodological critique of introspectionism, taking the line that mental processes could not be the object of scientific study because they were not open to observation. Watson's (1913) statement of position, "Psychology as a behaviorist views it," demanded a replacement of introspective method with the study of behavior. Partly, this was an impatient reaction to the irresolvably contradictory findings of the introspectionist psychologists. "Objectivity" was the catchword, and this meant focusing on events which both (a) could be reported reliably and were not susceptible to idiosyncrasy, and also (b) were open to observation by someone other than the person undergoing the experience. Watson recognized that this meant that psychology would no longer be the science of consciousness but he seems merely to have regarded this as a consequence of the requirement that psychology adopt a "scientific" methodology. It was not that consciousness was ill-formulated by the introspectionists, or that consciousness could be dismissed as unreal. It was simply not amenable to objective attack. It is also true to say that behaviorism was committed to the direct and unmediated connection between all human functioning and the world, to the extent that consciousness (which would seem to represent a hiatus in the flow of world—person exchange) was normally unrecognized.

This historical shift was unfortunate, for it put out of play several lines of thought which, when elaborated, are conducive to the development of qualitative and more specifically phenomenological psychology. When the psychologist concentrates on objective stimuli and measurable responses, attention is turned from the following (among other things):

The 'first person' perspective. Propositions about psychological events can only be stated in the third person—from the viewpoint of the observer rather than the actor themselves. The statement "*They* responded in such-and-such a way in certain environmental circumstances" may be scientific, but "*I* perceived (subjectively) the situation in such-and-such a way and so acted as I did" cannot be scientific.

The perceptual approach. Behaviorism could not consider the viewpoint of the research participant. And the other modes of intentionality of consciousness—thinking, judging, paying attention and switching it from one thing to another, etc.—could not be properly differentiated and researched because behaviorism could not permit itself to consider the relationship between consciousness and its objects of awareness.

Idiography. Behaviorist research, though allowing for 'individual differences' due to variations in individuals' histories of reinforcement, could not regard the study of people in their uniqueness as a justifiable scientific enterprise. Objectivity would be threatened.

Meaning is sacrificed by behaviorism. In the search for the objective and observable causes of behavior, the meaning that a situation has for the person disappears as a topic of research. Similarly, people's own accounts of their experience is regarded as *verbal behavior*—that is, responses which need to be explained in terms of their causes—rather than understandable and meaningful in their own terms.

Social relatedness was simply seen in stimulus-response terms: other people are an important source of stimuli, and my responses to them are likely to have significant repercussions. But people were not seen as different in kind to the things which constitute a person's environment; behaviorists were not able to recognize the *social nature* of the human being. In particular, they were not able to fully recognize the intersubjective constitution of human reality.

In effect, those things which behaviorism neglects provide a valuable list of items which are central to a qualitative sensibility in psychology. They also indicate the inimical context in which the general line of thinking of Husserl and his successors vied for a hearing.

However, within behaviorism, developments in a cognitive direction were made from time to time, attempting to re-establish psychology as a science of mental life (in some sense). Perhaps Miller, Gallanter, and Pribram's (1960) *Plans and the Structure of Behavior* is the most obvious of these, in that the authors termed themselves "cognitive behaviorists." Cognitive psychology can be seen as a critique of the behaviorist neglect of "inner processes," and the opening up of the possibility of studies of perception, memory, thinking and so on. But it retains a methodological commitment to observables; the novelty of cognitive psychology lay in developing models of inner processes on the basis of what was externally observable.

Both behaviorism and cognitive psychology became increasingly explicitly *positivist* in their view of psychology as a science. Positivism takes the view strongly that there is a real unitary world with definite characteristics; indeed, it is not even considered appropriate to construct an argument defending this view. The individual is part of this world, and so such processes as memory, emotion, thought, are events in the real world with definite enduring characteristics. The purpose of science is to set up experimental situations in which the characteristics of these psychological processes can reveal themselves, and this will allow the processes to be modeled. These models (mathematically formulated if possible) will show how certain variables interrelate, especially how they relate to each other in a cause-and-effect fashion. The purpose of research is to test hypotheses regarding relationships between variables, and to reach, by closer and closer approximation, theories which can begin to be regarded as having the status of scientific laws.

In rejecting positivism in this sense, qualitative psychologists influenced by phenomenology put on one side concern with some unequivocal real world, in favor of attending to the accounts that people formulate of *their* reality. It is this distance from the assumption of the primacy of realism that, very roughly, distinguishes qualitative research from behaviorism and allied cognitive psychology.

For some purposes, the radical behaviorism of Skinner (1904–1990) can be exempted from some criticisms of behaviorism in general, as Kvale and Grenness (1967) have shown. Emphatically, Skinner (1993) sought to develop an approach which was *anti-dualist*. There is to be no gulf between the mind and the material world. The problem of attempting to relate the external world to a quite distinct inner world of mental activity is regarded as an illusory one by Skinner.

> Suppose someone were to coat the occipital lobes of the brain with a special photographic emulsion which, when developed, yielded a reasonable copy of a current visual stimulus. In many quarters this would be regarded as a triumph in the physiology of vision. Yet nothing could be more disastrous, for we should have to start all over again and ask how the organism sees a picture in its occipital cortex. . . . It is most convenient, for both organism and psychophysiologist if the external world is never copied—if the world we know is simply the world around us. The same may be said of theories, according to which the brain interprets signals sent to it and in some sense reconstructs external stimuli. If the real world is, indeed, scrambled in transmission but later reconstructed in the brain, we must then start all over again and explain how the organism sees the reconstruction. (Skinner, 1964, p. 87)

Skinner did not simply set aside consciousness, as Watson seems to have done, but tried to deal with aspects of our "inner" life in terms of the behaviorist framework.

Asserting that the major factors which explain behavior are environmental ones, he redrew the "boundary" such that environment can be seen to include the body. The usual connotation of the word environment as suggesting "outer" in contrast to "inner" must be overcome. We are urged to remember that environment refers to the regions inside the skin as well as the body's surroundings. Feelings and thoughts are envisaged as behavior for Skinner.

Kvale and Grenness (1967) show that Skinner's non-dualist stance entails a rejection of the "illusion of the double world"—the dualism of an outer, objective, physical world and an inner, subjective, psychological copy. Most psychologists seem to retain this dualistic model, for instance in regarding conscious perception as an *interpretation* of physical sensations. This view is rejected by Skinner (1964, 1978). Another feature of the dualistic view is a "bifurcation of the public and private worlds." Skinner points out that there is a widespread assumption that a person has a special kind of access to their own "inner world" which is different from how they get knowledge of their surroundings (including other people). This idea is rejected by Skinner. He argues that we know the inner world through exactly the same processes that we know the outer world. The inner, "private" world is not different in kind from the outer, "public" world.

We shall see that much of the Skinnerian critique of cognitivism is paralleled by existential phenomenology. We act in the world, not in the head, and the meaningfulness of our world is directly and pre-reflectively available to us in perception. The difference between all kinds of behaviorism and phenomenological psychology lies in the commitment of phenomenologists to the point of view of consciousness (the world is of "my" meanings) rather than to the observer's point of view, and in the insistence that it can be that the origination of behavior is with the person as agent.

However, these considerations put us in danger of anachronism. Husserl's phenomenology was developing at the time of the fragmentation of psychology subsequent to the discrediting of introspectionism. Whatever *in principle* convergence or divergences there might subsequently have been between aspects of behaviorism and tendencies of existential phenomenology, this lay decades in the future and, in any case, by then the directions of psychology and of phenomenology were so far apart that few scholars, even if they were acquainted with the two discipline-areas, were able to undertake any comparisons.

HUSSERL'S PHENOMENOLOGY

Phenomenology as a Method for Establishing "Foundational Disciplines"

The founder of phenomenology as a philosophical movement, Edmund Husserl (1859–1938) had a fundamental aim, which it is necessary to have clearly in mind in assessing his work and its relevance for psychology. This aim was to provide a sure foundation for the different scholarly disciplines by establishing the meaning of their most basic concepts. This was to be done by a clarification of the essential structures of experience which distinguish one discipline from another and regulate the nature of each discipline's concepts.

A mathematician by training, we can discern in Husserl's work something of the approach to knowledge of Euclidean geometry; the laying out of axioms, on the basis of which things in reality can be thought through in a sound manner. Where this analogy breaks down is in the nature of the basic axioms, which are not, for Husserl, products of reason simply but the findings of the scrutiny of the abstract nature of the experience from which such-and-such a concept, fundamental to a branch of scholarship, derives.

Importantly, then, Husserl had a concern that the different sciences and scholarly disciplines had no methodology for establishing their basic concepts. Many psychologists with a taste for conceptual rigor also have a niggling concern that the basic concepts lack firm foundation and, if there are pretensions of such foundation, they seem nevertheless to be used loosely without an eye to any fundamental meaning. Husserl viewed this as typical of all the realms of scholarly thought. The aim of his original philosophical project was a rigorous one. Phenomenology was to be philosophy as a strict science (that is methodologically tight and producing indubitable results) and it would give a foundation to the concepts of each scholarly discipline. Thus there would (for example) be a phenomenological geography, effectively establishing what the discipline *is* by fixing its major concepts.

Husserl (1913/1983) argued that, for every empirical science there would be an *eidetic* discipline, a foundational body of phenomenological research that would not itself be directly concerned with the real world but would provide the set of well-established concepts which would allow the researcher to study the real world. "Eidetic" refers to the notion that these concepts would be of "essences"—the pure, foundational ideas giving rational structure to thought about reality. Just as logic is not specifically *about* reality, but provides the conceptual system allowing thought about aspects of reality to proceed, and just as geometry—an eidetic discipline—allows such technologies as land surveying to take place, so there would be an eidetic discipline for each realm of human endeavor.

> ... Positing of ... essences implies not the slightest positing of any individual factual existence; pure eidetic truths contain not the slightest assertion about matters of fact. (Husserl, 1913/1983, §4, p. 11)

So, just as geometry may posit a "straight line" as an item in its armory of eidetic truths—though it is another question entirely whether such a thing has any empirical manifestation—so the essences of other pure disciplines are to be established without attention being paid to the issue of *real existence.*

Among the phenomenological disciplines there would be a phenomenological psychology. In fact Husserl dealt specifically with this topic in the lectures in 1925 (Husserl, 1925/1977). But an account of a phenomenological psychology would be unlike any textbook of empirical psychology as usually understood, giving the results of scientific research. Rather it would examine and soundly found psychology's fundamentals so that empirical work could go ahead securely.

THE *THINGS THEMSELVES*—INTENTIONALITY AND THE PHENOMENON

How is the *experience of something* to be interrogated rigorously? It involves turning attention exclusively to experience. The possibility of turning attention to experience and to what is given in experience depends on the understanding that all consciousness is consciousness *of* something. This axiom, which Husserl heard from Brentano (1874/1995), entails the general characterization of consciousness as *intentional*, a term which refers to the idea that to be aware is necessarily to be aware of something. (A different meaning of intentionality to the usual one which refers to purpose or the aim of an action.) The first tenet of phenomenology is that consciousness is intentional. Yet Husserl (1913/1983) sounded a warning against this being misunderstood as simply reiterating the "inner" world and "outer" world dichotomy. Instead, it needs to be understood is that both the "mode of consciousness" and the "object of this consciousness" are *mine*, they are both within personal experience or awareness:

> ... [W]e all understand the expression "consciousness of something" ... It is so much more difficult to purely and correctly seize upon the ... peculiarities corresponding to it.... [N]othing is accomplished by saying and discerning that every objectivity relates to something objectivated, that every judging relates to something judged, etc.... For without having seized upon the peculiar own ness of the transcendental attitude and having actually appropriated the pure phenomenological basis, one may of course use the word, phenomenology; but one does not have the matter itself.... (Husserl, 1913/1983, §87 p. 211)

So "the phenomenon" is what appears (the intentional object) as described in its manner of appearing, with attention given to the conscious mode (perception, for example) in which it appears. We will consider later the two "aspects" of the phenomenon as described by Husserl (1913/1983), the *noema*, the *object* of awareness, and the *manner* in which one is aware of it, the *noesis*.

The key to the phenomenological approach is to steadfastly focus on "the appearance in its appearing." In stressing this, Husserl was being self-consciously anti-Kantian. To pay attention to the thing in its appearing is to dismiss the idea that there is a hidden *noumenon* lying behind the experienced phenomenon. No, the task is to describe what appears, the phenomenon, pure and simple. As Husserl tells us:

> ... [W]hat is decisive consists of the absolutely faithful description of what is actually present in phenomenological purity and in keeping at a distance all the interpretations transcending the given. (Husserl, 1913/1983, §90 p. 218)

Now, the description of one's awareness of the phenomenon is regarded by Husserl as indubitable. Thus, for instance, religious faith, though surely a matter of enormous dispute as to its relation to reality and its value in within society currently, is nevertheless describable in its appearing. It should be possible therefore, according to Husserl, for *anyone* to come to a statement of what faith is.

> No conceivable theory can make us err with respect to the principle of all principles that every origin presentive intuition is a legitimizing source of cognition, that everything originarily (so to speak, in its "personal" actuality) offered to us in "intuition" is to be accepted simply as what it is presented to us as being, but also only within the limits of what is presented there. (Husserl, 1913/1983, §24 p. 44)

THE PRIMACY OF "THE APPARENT" AND THE EPOCHĒ

It is no mean task to undertake a phenomenological description of what appears simply in its appearing. We have already implicitly accepted that "the phenomenon" is not a matter of *reality*. In turning attention to intentional consciousness, we have set aside concern with whether the object of awareness is part of some "outer reality." Husserl noted that this contrasts with ordinary life, in which we take it for granted that our activities and the mental life which is bound up with them refer to the real world. But in phenomenological work this assumption (the "natural attitude") has to be set aside:

> [T]he positing [involved in the natural attitude] undergoes a modification: while it in itself remains what it is, we, so to speak, "put it out of action" we "exclude it," we "parenthesize it." It is still there, like the parenthesized in the parentheses . . . (Husserl, 1913/1983, §31 p. 59)

The methodological move which is being described by Husserl here is the "stoppage" or epochē—"bracketing"—which neither denies nor asserts the reality of the phenomenon, but puts it out of play for descriptive purposes (Ashworth, 1996). Of course it is also worth noting that it may be a necessary part of the very meaning of a particular phenomenon for it to be understood to relate to reality (perception and its objects are like this), in such a case a relationship to reality is a part of the description of its appearing; it is not a pre-supposition but is found in the phenomenon itself.

If the assumption of reality must be bracketed, theories about the phenomenon must be similarly subjected to the epochē:

> The whole pre-discovered world posited in the natural attitude, actually found in experience and taken with perfect "freedom from theories" as it is actually experienced, as it clearly shows itself in the concatenations of experience, is now without validity for us; without being tested and also without being contested, it shall be parenthesized. In like manner all theories and sciences which relate to this world, no matter how well they may be grounded positivistically or otherwise, shall meet the same fate. (Husserl, 1913/1983, §32 p. 62)

The epochē is a central methodological concept of Husserl's phenomenology, then. Its sole purpose is to bring our attention fully to play on the thing itself as given in consciousness, shorn of any "surplus"pre-suppositions. It entails a setting aside of issues such as:

> The question of whether the 'thing experienced' is real or not. (It is not presumed to be unreal, or real, but this issue is set aside in order to attend to the experience itself.)
>
> Whatever previous opinion or scientific theory expects. (Thus a hypothetico-deductive approach is not used; rather the attempt is made to be exploratory as if *ab initio*.)
>
> Personal assumptions about the experience (e.g., its morality, its rationality, its coherence, its commonsense categorization). Attention is paid to the experience *for the experiencer* in as unprejudiced a manner as possible.

Husserl makes a very strong claim for the epochē, which, as we shall see, his existentialist critics found hard to accept:

> [W]e can... be assured of the legitimacy of the norm which we, as phenomenologists[,] intend to follow: To avail ourselves of nothing but what we can make essentially evident by observing consciousness itself, in its pure immanence. (Husserl, 1913/1983, §59 p. 136)

Essences and the Intuition of Essences

Husserl did not regard phenomenology as staying simply with the description of specific phenomena as given in experience. Rather—in line with his aim of developing a set of non-empirical disciplines each of which serves to specify the conceptual structure of some scholarly discipline or other—he laid our a further methodological move which would reveal the *essence* of a phenomenon. Beyond describing a certain phenomenon "in its appearing," Husserl wished, further, to give an account of its *essential features*. Here we have generalization of a certain sort. Not generalization from a sample to a population, but a movement from the particular description to the description of a total class. One could say that the essence refers to that which is example we proffer is an example *of* (cf. Cumming, 1992).

Husserl suggested procedures for securing the essence of a phenomenon such as "free imaginative variation," in which a particular example of a phenomenon would be taken, and the question would be put, If such-and-such a feature of the phenomenon, as described on the basis of the examples we have used, were removed or altered in some way, would we still regard it as an instance of that phenomenon? If so, the aspect remains as part of the essence (or eidos) of the phenomenon. Here we have a second, eidetic epochē. Phenomenology is defined, for Husserl, in the following statement:

> As for phenomenology, it is concerned to be a *descriptive* eidetic doctrine of transcendentally pure mental processes as viewed in the phenomenological attitude; and... it has its inherent legitimacy. Whatever can be apprehended eidetically in

> pure intuition as belonging to reduced mental processes, either as a really inherent component part or as an intentional correlate of the latter, properly belongs to phenomenology... (Husserl, 1913/1983, §75 p. 167)

Plainly, this dense paragraph needs to be pulled apart if its import is to be clear. Phenomenology is now being defined by Husserl as the discipline leading to a set of statements concerning the precise manner in which things appear in consciousness. But he emphasizes that the focus is on what is *essential* about these ways of appearing. Our experience (or, better, because it is Husserl's term for the specific mode of cognition employed in "seeing" essences, our intuition) of such ways of appearing, just as they appear, will set out the essentials of the mental acts in which they appear (the essence of perception, for instance) but also, in sofar as there are invariant features of the objects of these mental acts, will describe these features of the objects of awareness, too. Intuition here is being used to refer to the apprehension in awareness of the phenomenon and its essential features. The essence will be in some sense a universal. Take perception, for example:

> Thus we describe and, in so doing, determine by *strict* concepts the generic essence of perception taken universally or that of subordinate species such as the perception of physical things... Prior to these, however, are the highest universalities: the mental processes taken universally, the cogitation taken universally, which already make extensive essential descriptions possible. (Husserl, 1913/1983, §75 p. 168)

The highest level of universality, here, will be intentional consciousness itself. It is clear that Husserl wished for eidetic phenomenology that it would describe the conceptual realm of the different scholarly disciplines and also provide universal concepts which would underpin thought generally. In *Ideas 1* (Husserl, 1913/1983) the volume on which much of the exposition in this chapter has been based (partly because it was this phase of Husserl's thought against which the existentialists, following Heidegger, rebelled) Husserl did develop an eidetic account of some areas of logic and taxonomy. In the second volume, *Ideas II*, (Husserl, 1952/1989) which, in the German original, was actually published posthumously, we have account of the constitution of such fundamental realms as material nature and animal nature (including "the psychic"). These high-level accounts may be termed, 'regional ontologies', descriptions of the essential characteristics of particular "kinds" of being.

Despite these successes, it is worth noting that Husserl himself seems to have been concerned about the epistemological status of eidetic intuition. He sometimes presented it as giving rise to self-evident results, ones which were just as self-evident as the initial description of the phenomenon in its appearing. But the following passage indicates a hesitation. Essences do not have to be validated against reality. Nevertheless, they depend on a certain kind of self-observation which is open to critique:

> As sciences of pure essence, geometry and phenomenology do not recognize any findings about real existence.... If, now, phenomenology does not... have to make

> existential findings about mental processes, if it thus need not make "experiences" and "observations" in the ... sense in which a science of matters of fact must support itself by them, it nevertheless makes eidetic findings ... [which] it owes ... to reflection, more precisely to reflectional intuition of essences. As a result the skeptical doubt with respect to self-observation likewise comes into view for phenomenology; it comes into view for phenomenology, more particularly, in so far as this doubt allows of being extended ... from reflection on something immanent to reflection taken universally. (Husserl, 1913/1983, §79 p. 184)

THE *NOESIS* AND ITS *NOEMA*—THE MODE OF CONSCIOUSNESS AND ITS INTENTIONAL OBJECT

It was noted earlier that Husserl was at pains to emphasize that intentionality did not mean, for him, that consciousness (as an "inner" power) was always related to an "outer," real object. Rather, intentionality related consciousness to an "immanent" object, an object of awareness, which might or might not have some kind of reality beyond its mental "presence." In *Ideas I* and *Ideas II*, the founder of phenomenology made especially sure that the inner/outer interpretation would be ruled out of court by naming the mental act which constitutes a phenomenon the *noesis*, and the intentional object the *noema*.

> We have bestowed such great care on working out universally the difference between noesis (i.e., the concretely intentive mental process ...) and noema because the seizing upon and mastering it are of the greatest importance for phenomenology, and indeed decisive for the legitimate grounding of phenomenology. At first glance it would seem to be something obvious: Any consciousness is a consciousness of something, and modes of consciousness are highly diversified. On approaching more closely, however, we become sensible of the great difficulties involved. They concern our understanding of the mode of being of the noema, the way in which it is "implicit" in the mental process, in which it is "intended to" in the mental process. Quite particularly they concern the clean separation of those things which, as its really inherent components, belong to the mental process itself and those which belong to the noema ... (Husserl, 1913/1983, §96 pp. 233, 234)

We see in this paragraph, also, a dilemma that was beginning to surface. Whereas the "mental act" (perceiving, judging, and so on) could be regarded as definitely essential to the phenomenon—"inherent" to its appearing in awareness, the noema was more problematical. Since noemata are not inherent, self-sufficient, characterizable in their eidos—they are not as such part of the essence of the phenomenon. But a footnote indicates that Husserl was not confident about the characterization of the noema as excluded from the phenomenon.

> ... [T]he Eidos of the noema points to the Eidos of the noetic consciousness; both belong together *eidetically*. The intentive as intentive is what it is as the intentiveness belonging to consciousness *structured* thus and so, consciousness which is consciousness of it.

> In spite of this non-selfsufficiency the noema allows for being considered by itself, compared with other noemas, explored with respect to possible transformations, etc. (Husserl, 1913/1983, §98, p. 241)

There is a "problem" of the noema—its multiplicity and refractoriness to eidetic specification. Husserl regarded the noesis—the mental act constituting the phenomenon as more interesting and more productive for his philosophical purpose. Judging, then, was more focal for him than was the unlimited set of judgments. And yet:

> ... [E]idetic description of consciousness leads back to that of what is intended to in it, that the correlate of consciousness is inseparable from consciousness and yet is not really inherent in it. The noematic became distinguished as an *objectivity* belonging to consciousness and yet specifically peculiar. (Husserl, 1913/1983, §128, p. 307)

In the end phenomenology must describe both "poles" of the intentional correlation between noema and noesis:

> ... [A] systematic phenomenology is not allowed to direct its aim one-sidedly at an analysis of what is really inherent in mental processes and specifically of intentive mental processes. (Husserl, 1913/1983, §128, p. 308)

THE EXISTENTIALIST CRITIQUE OF HUSSERL'S PHENOMENOLOGY

Husserl's project came under criticism especially from his own students. Indeed, phenomenology has over the last 90 years shown itself especially fissile. Heidegger, a favored student of Husserl, provided the most telling critique in *Being and Time* (1927/1962). In it he drew together, with almost inconceivable creativity, Husserl and Kierkegaard—two writers at almost opposite intellectual poles. Husserl the seeker after rigor and conceptual foundations; Kierkegaard the early advocate of an anti-foundational (almost postmodern) position, a thoroughgoing existentialist. And Heidegger, in criticizing essentialism, also built on the hermeneutic tradition to develop a philosophy which was not grounded, but which involved an interrogation of what (seemingly) *is*.

How did Heidegger's existential and hermeneutic phenomenology work? It entailed the rejection of Husserl's view that the detached philosopher could interrogate experience and emerge with universally-valid accounts of the essential structures of the sciences and scholarly disciplines. It also involved an assertion that the person is inescapably part of the historical-cultural world. (Being-in-the-world is a technical term for the situation of the human being.) Therefore—at least in the earlier writings—he thought it was necessary to investigate Being-in-the-world before turning attention to anything else. Husserl castigated this as a mere anthropology, a line of work appropriate to scientific research not philosophy.

How is human Being-in-the-world to be investigated as a phenomenon (for Heidegger still saw himself as doing phenomenology even if Husserl's method was adjusted and new aims put in place)? It was to be interpretative, a hermeneutics. It was to take due account of the importance of language. The collective, historical, cultural position of the individual person was to be regarded as of great weight. And, although phenomenology was no longer seen as a rigorous discipline aiming to give solid conceptual foundations to the sciences and humanities, Husserlian thinking still shines through in Heidegger, even if the kind of approach has changed. We do get certain essential structures of Being-in-the-world, though they might be in a different key to Husserl. It could be said that the work of Heidegger has, to a psychologist, a less cognitive flavor than Husserl (who can sometimes seem to be describing "mental mechanisms").

Phenomenology as Hermeneutics or as a Descriptive (Elucidatory) Discipline

Heidegger stresses hermeneutics. But what is this to mean for phenomenological psychology? There is much debate about the role of interpretation and whether it contrasts with the attempt "simply to describe," or elucidate, experience. As Palmer (1969) points out, hermeneutics has in the last 200 years or so been applied with very great generality to the process of coming to an understanding:

> . . . [S]omething foreign, strange, separated in time, or experience, is made familiar, present, comprehensible; something requiring representation, explanation or translation is somehow 'brought to understanding'—is 'interpreted' (Palmer, 1969, p. 14)

Take these examples:

When J.S. Bach wrote at the head of his scores *Ad maiorem Dei gloriam* was he correct in his account of his motivation? Informed by the interpretations of such social thinkers as Marx, we might conceivably think that Bach was ill-informed about the true basis of his work, which is actually to be understood as inextricable from the power structures of his day; he earned his living by aggrandizing the local nobility. (In fact there is evidence that this is not true of Bach, but the illustration makes the point.)

Or when Beethoven inscribed a short piano piece, *Für Therese* (or, as his publisher had it, *Für Elise*), was he right? Was it an expression of his adoration of Therese? Again, educated by Freud, we might today believe that, underlying this ostensible motive, was a structure of unconscious desire which, while it might at that moment have had Therese as its focus, would be better understood by an interrogation of the unconscious.

In these instances, Freud and Marx both provide an interpretive approach which would aim to go beyond the immediate understanding of Bach and Beethoven's motives to provide a "truer" view. And other less elaborate theories of society or motivational psychology have this function. We may typify these analyses in terms of Ricoeur's (Ricoeur, 1970; Robinson, 1995; Smith, 1995) description of the hermeneutics of suspicion which reaches, beneath the research participants' accounts, a "truer" understanding drawing on a notion that the accounts may entail certain taken-for-granted assumptions which are not explicitly stated but which, when noticed by the researcher, can be used to shed light on what is said so as to render it more meaningful.

> *The 'enriching' mode of interpretation* (Ricoeur called this the 'hermeneutics of meaning-recollection'). This aims at a greater understanding of the thing being analyzed, but always a greater understanding *in its own terms*.
>
> *The hermeneutics of suspicion* finds that, behind the thing being analyzed, there is a further reality which allows a much deeper interpretation to be made and which can challenge the surface account. It involves a willingness to suspect the surface account and to question rigorously.

In my view, interpretation of a sort is involved in describing experience phenomenologically. But it is one that aims at the experience in its own terms—it is the hermeneutics of meaning-recollection, the effort at *elucidation*. Now in Bach or Beethoven's experience, issues of class and desire could perhaps be found. But these categories are not to be presupposed. They may be described if and only if they are intrinsic to the experiences for Bach and for Beethoven.

The Person in Heidegger's Phenomenology

Dasein seems to be synonymous with Being-in-the-world in Heidegger as a term for beings of the human kind. The term carries no connotation of personality. What he emphasizes is that reflection is possible for us-—in fact, we are necessarily interpretative beings, making sense of the world in which we find ourselves, a world of spatiality, temporality, sociality and discourse.

Heidegger does not use the term "lifeworld" which will become important soon in this paper; he just uses the word "world." The words are synonymous. The world *is* my lifeworld; the world is seen from my perspective.

In describing the human being, Dasein, Being-in-the-world, Heidegger draws from Kierkegaard the view that a person is permanently anxious about their own "existence," that is, their own meaning in the world. Of course Laing (1965) made much of this. But less extravagantly (and maybe more directly descriptive of actual experience) Heidegger described the connection between Dasein and their world as *care*. This term is not intended to carry the heavy connotation of emotional solicitude; rather it is simply that the things of our world *matter* to us in the way that its sensed-and-acted-in environment does not matter to a machine. Machines do not have a world in the human sense.

Sartre, de Beauvoir, and Merleau-Ponty

The spread of existential phenomenology in the middle decades of the twentieth century was especially due to Sartre (1943/1956) and de Beauvoir (1949/1988) who, for the most part, took a neo-Heideggerian view. The main distinction from Heidegger was that Sartre can be regarded as running dangerously close to dualism by imposing a radical distinction between consciousness and the objects of consciousness, which Heidegger did everything in his power to avoid. For Sartre, all that can be thought, perceived, imagined is an object of consciousness. The body is an object of consciousness; so is the self. In this he was arguably following a line of thought due to Husserl:

> The theory of categories must start entirely from this most radical of all ontological distinctions—being *as consciousness* and being as something which becomes *"manifested"* in consciousness, "transcendent" being—which, as we see, can be attained in its purity and appreciated only by the method of the phenomenological reduction. In the essential relationship between *transcendental* and *transcendent* being are rooted all the relationships... between phenomenology and the other sciences... (Husserl, 1913/1983, §76, pp. 171, 172)

Consciousness (like Heidegger's Dasein) has no intrinsic characteristics. It is free. That is, within the situation, as consciousnesses we can imagine alternatives. Self as an object of consciousness is (effectively) imaginary, and in identifying with a self, we adopt it. The possibility of having a self is due to our membership of society for it is in interaction with others that we take an external view on our actions and bodily presence, and are thus enabled to form a self. To believe in the *solidity* of our self as if it was *not* open to free alteration (again, within the limits of the situation) is "bad faith," self-deception (Sartre, 1943/1956).

> Doubts, remorse, the so-called 'mental crises of conscience', etc.—in short, all the content of intimate diaries—become sheer *performance*. (Sartre, 1937/1957, p. 94)

So here we have a non-essentialist account of selfhood. In this, Sartre (1937/1957) claimed to be correcting Husserl (and specifically the Husserl of *Ideas I*), but this is not the case:

> In these peculiar combinations with all of "its" mental processes, the Ego living in mental processes... is not something taken *for itself* and which can be made into an Object *proper* of an investigation. Aside from its "modes of relation" or "modes of comportment" the Ego is completely empty of essence-components, has no explicable content, is undescribable in and for itself: it is a pure Ego and nothing more. (Husserl, 1913/1983, §80, p. 190)

Despite the fact that we can never *coincide* with the self which we adopt, Sartre argued that, in an attempt to allay the anxiety of freedom, we do identify ourselves as someone with defined characteristics and *unnecessarily* limited choice (even give the limitations of the situation). For Sartre there is a further dynamic involved in the choice of selfhood. He argues that each self has a theme, consistent throughout the biography, which indicates an "original choice" or personal *project* of what "kind" of person to be. Examples of the interpretation of such original projects are found in Sartre's biographical works.

I think it is right to say that Sartre and de Beauvoir's existentialism goes beyond the confines of careful phenomenological description, notably when they speculate about original choice and project (the work they refer to as "existential psychoanalysis," thereby pointing implicitly to the fact that they have moved into the hermeneutics of suspicion). At other points they do go beyond the immediately apparent—and move outside the realm of phenomenology proper—as when Sartre distinguishes so firmly between consciousness and its objects even though we can of course never really come to an awareness of consciousness without catching it directed to an object of awareness (the rule of intentionality indicates their inextricability). But here I would argue that the distinction is of heuristic value to phenomenology, as is Husserl's noema/noesis divide.

Maurice Merleau-Ponty's (1945/1962) existential phenomenology is the version which underlies the work of the majority of phenomenological psychologists today. He is silent on some of the more extravagant Kierkegaardian themes of the other existentialists. He maintains some commitment to Husserl's epochē as a technique of thought, but takes the Husserlian aim of discovering incontrovertible truths using the phenomenological method as a vain hope (because the philosopher, like everyone else, is immersed in a social, historical world). Nevertheless, the return to lived experience, carefully elucidated, provides an approach to elucidating the basics of the realms which science investigates. The concepts of science are secondary elaborations upon experience, which is primary.

In Merleau-Ponty, the danger of dualism in the work of Sartre is combated by emphasizing the rejection of the image of the person as a detached consciousness. The person is a "body subject," ambiguous in its conscious-materiality. Intentionality is the characteristic of consciousness—but it is a characteristic of the body subject more generally, as when, unreflectively, I drive in a way which is precisely attuned to the conditions of the road despite lack of what we might call the "reportability" of my body's decisions.

As we have seen, for existential phenomenology the analysis of experience is not intended primarily to provide a foundation for basic scholarly concepts but to describe the varieties of human engagement. Husserl's (1936/1970) late elaboration of the notion of the lifeworld was still in the service of developing indubitable concepts for psychology and the other disciplines, but it was taken in an existentialist direction by such authors as Heidegger, Merleau-Ponty and Sartre.

THE LEGACY OF HUSSERL FOR PSYCHOLOGY

Husserl established that human experience in general is *not* a matter of lawful response to the "variables" that are apparently in operation. Rather, experience is of a meaningful Gestalt wrapped up in the lifeworld. In other words, the human realm entails embodied, conscious relatedness to a world. The natural scientific approach (in terms of discrete variables acting in a causal way so as to produce lawful effects in human experience and behavior) is inappropriate. Human meanings are the key to the study of lived experience, not causal variables.

The focus of qualitative psychology based on the legacy of Husserl is on lived experience and the attempt to describe some particular feature of it to the scientific community in a clear way that will inform understanding. It is empirical, rather than being based on purely theoretical reflection. It does not follow Husserl in his idealist attempt to develop a full system of clarified concepts on which psychological research can be based, though some empirical work may develop in that direction. Rather, it describes a certain lifeworld or some phenomenon within the lifeworld.

THE EARLY CONFLUENCE WITH GESTALT PSYCHOLOGY

Gestalt psychology dissented especially from the *atomism* of Wundt. The emphasis on the characteristic of "total form quality" of mental life, which is embodied in the very name of this school of psychology, is its distinctive feature. Some of the most representative findings of Gestalt psychology in the field of perception are described in Fig. 2.1.

The phenomenological aspect of the Gestalt approach lies in its focus on the experience as such. Thus, it would be unphenomenological to describe apparent movement by saying that the person "brings the elements together" to see movement. Rather, to the experiencer, what is perceived *is* movement (there is no "apparent" about it), and no personal activity is experienced as being involved in producing the perception. Moreover, the refusal to atomize experience by referring to distinct "variables" is embodied in the Gestaltists' basic postulate, "The whole cannot be accounted for by the sum of the parts."

One temptation which Gestalt psychologist wrestled with (see Gurwitsch, 1964) was to view experience as the lawful outcome of cognitive processes. Thus laws of perceptual form—such as *continuity*, which enables us to see that it is a baby behind the cot bars and *closure*, which enables us to see that the interrupted lines of the Sheffield Wednesday Football Club motif represent an owl—pose a problem. Are they part of the stimulus that impinges on the eyes, or are they aspects of a process of decoding in the nervous system? The presupposition with which they had to wrestle was the cognitive one which assumes that we, primarily, have raw sense data and to this is added interpretative information which turns it from sensation to perception. In the end, the Gestalt theory set the cognitive model, the "constancy hypothesis," aside in favor of a description of the features of the experience itself. In doing so they, incidentally, made the account more phenomenological.

> It is no longer possible nor even necessary to distinguish between those features of perception which are genuine sense-data and those which are considered to be contributed by [other] sources . . . —a distinction which, as Köhler pointed out, is based on physiological considerations. *By the dismissal of the constancy-hypothesis and*

FIGURE 2.1. Gestalt qualities of perception.

This picture illustrates a number of Gestalt perceptual laws, the most general of which is *prägnanz* ("good form"), the principle that experience is as clear and meaningful as possible. More specifically, Gestalt psychologists note that the principle of *proximity* "invites" us to see items which are relatively close to each other as a grouping. Also parts which are "missing" from the stimulus array are experienced as being present—there is a law of *closure*. A primary law of all perceptual arrays is discrimination of *figure* as distinct from *ground*. To see the cat is to relegate other items of the picture to the background.

It is possible to specify over a hundred such laws of perceptual form.

A picture which provides such poor information as the one on this page makes us more aware of Gestalt principles, perhaps. But it is important to realise that what is true here is as true of all perception, not just the perception of *obviously* ambiguous or taxing arrays. To sit by the side of a cot and see the baby as a baby, means that the visual form of the baby is not broken by the imposition of the vertical cot bars. An equally effortless "achievement" is to see something as being distant rather than small. A main difficulty for those working on the machine detection of specific patterns of light, is that all perceptual fields are in principle ambiguous.

Gestalt laws do not just relate to static experience, nor just to the visual world. Actually, the earliest work of one of the three leading Gestalt theorists, Max Wertheimer, had to do with the study of apparent movement—the fact, well-known to the designers of illuminated advertising signs, that a number of lights in close proximity to each other which are switched on and off in sequence will invoke the experience of motion.

> *the ensuing abandonment of the above distinction, the descriptive orientation with regard to perception is reinstated in its own right. All features displayed by perception must be treated on the same footing*. (Gurwitsch, 1964; emphasis in original)

The phenomenologically-relevant contribution of the Gestalt school to psychology was to stress the importance of close interrogation of experience *as experience* without being led astray by assumptions about what experience "should" be (e.g., it should be an outcome of certain psychophysiological processes, or it should be compounded of certain basic elements). An epochē is involved.

Heider and Commonsense Psychology

Heider (1958) is the major source of "attribution theory" as a line of research in social psychology. In accord with the program of psychological phenomenology, Heider set out to describe the features of the commonsense attribution of causes of events. What state of the experiential field is it that leads a person, observing an event, to see it as being the outcome (for instance) of the deliberate, motivated action of someone—or an accident? This is an area of "commonsense psychology," the study of the implicit theories of human experience and behavior that individuals actually use in daily life.

Heider's method of research seems not to be empirical in the usual sense. Nor is it introspective in the manner of the nineteenth-century psychologists. Reflection on everyday occurrences and personal experience of making causal attributions furnished his descriptions. His scholarly origin in Gestalt theory and acquaintance with phenomenology allow us to see how he was able to work against the hugely dominant current of behaviorist thought.

Lewin and the Personal Motivational Field

Both Heider and Kurt Lewin were in origin Gestalt psychologists, and were acquainted with the idea that the "mental field" can be viewed as a field of forces, for, beyond the description of the phenomenology of perception and other psychological processes, Gestalt theorists saw the explanation of mental phenomena in fields of cortical arousal which would map, in a complex way, onto the experience. Using this language of forces within a force-field, Lewin (1935, 1936) developed a way of presenting the plethora of factors influencing a person *motivationally*. What a person would do in a given circumstance could be portrayed as an outcome of positive and negative forces acting in various directions in relation to possible lines of action.

Lewin's work in mapping the subjective pressures and tendencies which are experienced in the context of making choices or succumbing to circumstances leads directly to important applied endeavors in the areas of group processes (including the development of so called T-groups, and the delineation of modes of leadership) and "action research." The situation for a group can be viewed in similar terms to the situation for an individual, and so the dilemmas in group conflict can be mapped out and the alternatives considered. In doing action research, the organizational expert (it might be anyone from a classroom teacher to a management consultant) appraises the situation in which an individual or group finds themselves, and decides on a way of altering the situation or, more indirectly, facilitating the alteration of the situation by those involved. The consequent change in the organizational field is observed (using any relevant techniques but usually in the context of participant observation) and the need for further changes considered—and so on.

AMERICAN "PSYCHOLOGICAL PHENOMENOLOGY"

In the generation subsequent to the advent of behaviorist psychology, a non-behaviorist and non-psychoanalytic line of thinking developed in the United States. This was a loose tradition, but self-conscious—and especially aware of its position in contrast to the dominance of behaviorism in the academy and psychoanalysis in the consulting room.

GORDON ALLPORT: AN IDIOGRAPHIC APPROACH

Although insufficient attention has been paid by qualitative psychologists generally to the work of Gordon Willard Allport (1897–1967), he has not been entirely unnoticed (e.g., Smith, Harré, & Langenhove, 1995). Allport sometimes took a harsh line regarding mainstream psychology. In 1965, he told the American Psychological Association that experimental psychology

> dashes forth like a headless horseman. It has no rational objective. It uses no rational method other than the mathematical, reaches no rational conclusion. It lets the discordant data sing for themselves. (*New York Times* obituary. 9th October, 1967)

However, the criticism was not simply directed against quantitative empiricism (he himself had published influential studies of this kind) but rather against its dominance, and the lack of appreciation of other ways of researching the human person. Allport developed a theory of personality that stressed the uniqueness of the individual.

> I object strongly . . . to a point of view that is current in psychology. Eysenck states it as follows: To the scientist, the unique individual is simply the point of intersection of a number of quantitative variables.
>
> What does this statement mean? It means that the scientist is not interested in the mutual interdependence of part-systems within the whole system of personality . . . [and] is not interested in the manner in which your introversion interacts with your other traits, with your values, and with your life plans. The scientist, according to this view, then, isn't interested in the personality system at all, but only in the common dimensions. The person is left as mere "point of intersection" with no internal structure, coherence or animation. I cannot agree with this view. (Allport, 1961, p. 8; his italics)

We can see here a reason why Gestalt theory appealed to Allport. His view is closely reminiscent of the principle that any whole entity—in this case, the whole person—has to be understood as a coherent structure, not simply as a collection of elements. So, in this statement, the holism of Allport's approach is plain, and his interest in the *idiographic* approach to psychological research logically follows. The individual may be studied as a unique case. The psychology of personality does not need to be exclusively *nomothetic*, restricting its attention to general dimensions on which individuals vary. The nomothetic approach assumes that the behavior of a particular person is the outcome of laws that apply to all, and the aim of science is to reveal these general laws. The idiographic approach would, in contrast, focus on the interplay of factors which may be quite specific to the individual. It may be that the factors only take their specific form in this person; certainly they are uniquely patterned in the given person's life (Allport, 1962). There is a sense in which no two people can have the same trait of personality because its coexistence with other traits and personal circumstances will materially affect its nature.

As tends to be the case with psychologists who take a humanistic line, Allport considers in some depth the meaning of the *self*. He tries to cover a very great deal of what is linked in ordinary language in some way or other to the notion of self, using the coinage "proprium"—including both the individual's *conception* of the self and the aspects of their world which they may be said to identify with, and the "integrative" mental function which may be labeled "selfhood." He goes on to draw the distinction used by both James and Mead for, I think, different purposes. Part of the problem in formulating a view of the self (and this is a problem in other areas of theory-formation for Allport) may be the lack of clarity concerning the distinction between a third-person, external view of the person—the person as observed by the psychologist—and a first-person, internal view—the person they themselves perceive or construe.

Despite his insistence on the importance of the idiographic approach, this does not seem to imply for Allport a necessary emphasis on qualitative methods of research. His recommended approach would be to study an individual person by using as many and varied means as possible. But he did pioneer some interesting qualitative approaches, such as the analysis of "personal documents" and the use of "self reports" as a means of understanding the individual (Allport, 1961, pp. 401–414, 1965).

The viewpoint of Gordon Allport is hard to epitomize without distortion because of his eclecticism. But his espousal of qualitative methods in psychology largely reflects his concern with the individual as a totality within their world of experience. This holistic and idiographic concern has, as he recognized, affinities with Gestalt theory and with existentialism. But—despite allowing self-report as a research technique—Allport does not seem to have been interested in a qualitative psychology from the viewpoint of the person themselves. In the end, his is a psychology which will describe the person in their individual complexity, yes, but carried out from an external vantage point by the psychologist with personal documents and other "subjective" material to be used as evidence.

Robert MacLeod

In the 1940s and 50s, a number of American psychologists, boosted by a stream of immigrant European scientists and philosophers escaping Nazism, began to write of a "psychological phenomenology." This literature did not stress the philosophical basis of its approach, though the writers were generally well-informed about the phenomenology of Husserl. Rather, the stress was on the need for psychology to research the personal "field of experience" of individuals. It is notable that the indirect effect of phenomenological philosophy through Gestalt theory was as important to American psychological phenomenology as was direct Husserlian influence, and Gestaltists were among the immigrants.

The leading figure in this movement was Robert MacLeod, among whose publications was an apparently well-received paper published in the leading *Psychological Review* (1947). Psychological phenomenology involves, MacLeod tells us, the adoption of

> an attitude of disciplined naïveté. It requires the deliberate suspension of all implicit and explicit assumptions, e.g., as to eliciting stimulus or underlying mechanism, which might bias our observation. The phenomenological question is simply "What is there?," without regard to Why, Whence or Wherefore.

In exemplifying some of the assumptions which require such suspension in order to carry out phenomenological description, he pays particular attention to various "biases," which are usual and taken for granted.

> *The atomistic-reductive bias*—by which is meant a concern with the elements of mental world, the simple sensations, feelings and ideas that, it was supposed, combine in a mental chemistry to provide the thinker/perceiver with the complexity of actual experience (he is harking back here to Wundt).

> *The stimulus-receptor bias*—'Granted the search for elements, it is understandable that experience was considered elementary which seemed to be the direct result of the activation of a simple receptor mechanism. By the same logic, any experience which evinced no such correlation was accorded a secondary status.' (p. 195)

> *The genetic bias* is the search for origins and the assumption that the earliest form of a psychological fact is truest and gives greatest understanding.

Without making heavy weather of it, MacLeod draws his readers into a specification of those pre-suppositions which are usual in social psychology which must be bracketed in order to attend to social psychological phenomena "in their appearing." Among the areas of social psychological interest he discusses is the self. He points out that there is a sense of self in which "self" is part of the field of experience—an object of awareness. He merely alludes to the contrasting meaning of self as the assumed entity which is not *in* experience but is taken to be the being which "has" the experience (so MacLeod points to, but does not elaborate, the Jamesian distinction between the self as "knower" and as the "known"). What he wants to emphasize is the phenomenological requirement that we see "subjectivities" such as prejudiced attitudes, not as characteristics of a self, but as states of the experiential field, which are to be understood in that context.

MacLeod discussed, though in passing, the question of the techniques of research which might be the most appropriate ones for the phenomenological study of psychological worlds, commending approaches which are "less constrained by conventional categories" such as "a variant of the free interview method" (p. 207). Later, he stresses the need for "improved methods of observation" (p. 208):

> In this connection the products of casual, uncontrolled observation must not be scorned.... [M]uch can be gained from the persistent attempt to note and describe the social phenomena which occur incidentally in one's own experience. It is notoriously difficult, however, to maintain a disciplined naïveté about oneself, and ... the social psychologist must improve his methods of studying the social field of the other person.... [I]t is suggested that the most promising method is that of the intensive interview.

Snygg and Combs

Two other authors within this tradition, who deserve mention are Donald Snygg and Arthur W Combs (Combs & Snygg, 1959; Snygg & Combs, 1949) who tell us that

> Human behavior may be observed from at least two very broad frames of reference: from the point of view of an outsider, or from the point of view of the behaver himself.... [T]he second approach ... attempts to understand the behavior of the individual in terms of how things 'seem' to him. This frame of reference has been called the 'perceptual,' 'personal,' or 'phenomenological' frame of reference and is the point of view of this book. (Combs & Snygg, 1959, p. 16)

In fact, these terms labeling the first-person viewpoint do seem to be synonymous to Snygg and Combs, for the main systematic alteration between their first and second editions is the replacement of "phenomenological" with "perceptual" throughout the book. In terminology reminiscent of Lewin, they employ the notion of field.

> ... [W]e shall use the field concept to refer to that more or less fluid organization of meanings existing for every individual at any instant. We call it the perceptual or phenomenal field. *By the perceptual field, we mean the entire universe, including himself, as it is experienced by the individual at the instant of action.* It is each individual's personal and unique field of awareness, the field of perception responsible for his every behavior (p. 20, their italics)

Snygg and Combs take the question of research approach seriously (the final chapter is devoted to it), and their main concern is that understanding another person depends on knowing how they perceive themselvs and their world. They admit that this depends on knowledge "not open to direct observation by an outsider" (p. 439). But later they suggest a combination of such observations, information from the person themselves—including diaries, informal conversation, autobiography and self reports—projective techniques and therapeutic records.

The "Humanistic Psychologists"

Especially in the American context, phenomenology and existentialism have been linked with "humanistic psychology" (see Misiak & Sexton, 1973). In the 1950s and 1960s, psychological thinking in North America was dominated by behaviorism and psychoanalysis. The determinism shared by those approaches was among the features which led a number of authors, notably Allport (1961, 1962, 1965), Bühler (1971), Maslow (1968) and Rogers (1967)—many of whom were psychotherapists—to call for a "third force" of psychological thinking (Bugental, 1964) to counter the reductive tendencies of the contemporary psychological mainstream.

Humanistic psychologists are a very diverse group, but the characteristics officially listed (Misiak & Sexton, 1973, p. 116) are:

> A centering of attention on the experiencing person and thus a focus on experience as the primary phenomenon in the study of [the individual]. Both theoretical explanations and overt behavior are considered secondary to experience itself and to its meaning to the person.
>
> An emphasis on such distinctively human qualities as choice, creativity, valuation, and self-realization, as opposed to thinking about human beings in mechanistic and reductionistic terms.
>
> An allegiance to meaningfulness in the selection of problems for study and of research procedures, and an opposition to primary emphasis on objectivity at the expense of significance.
>
> An ultimate concern with and valuing of the dignity and worth of [the individual] and an interest in the development of the potential inherent in every person. Central in this view is the person as [they] develop [their] own being and relate to other persons and to social groups.

Although most humanistic psychologists express their approval of phenomenology and of existentialism (some textbooks of personality theory treat humanistic psychology and phenomenology as virtually synonymous), this association seems mainly to be due to the concern with consciousness in the methodology of the former and the appearance in existentialism of concepts such as authenticity, freedom and personal project. There seems to be less commitment to the methodological rigor of phenomenology or the capacity of existentialism to explore the human possibilities of imagination and creativity as part-and-parcel of other aspects of the human condition such as anxiety and self-deception. However, this sweeping opinion does not, I admit, do justice to the variety of the school.

CONTEMPORARY PHENOMENOLOGICAL PSYCHOLOGY

With Amedeo Giorgi and Max van Manen we reach contemporary phenomenological psychology in its extreme versions. Both are explications of the lifeworld. Giorgi's (1970, 1985) is relatively tight methodologically and has something like "essences" in view, aiming by empirical work to arrive at the essential features of such things as anger, loneliness, and different modes of learning (see Table 2.1).

TABLE 2.1. Giorgi's husserlianism (Giorgi, 1985).

1. Descriptions of concrete experiences, supposed by the research participant to be an example of the matter of interest to the researcher, are elicited by interview or by asking for written accounts.
2. The transcription or account is subjected to "meaning unit analysis"—the text is divided into (possibly overlapping) sections each of which indicates a distinguishable meaning related to the phenomenon under study.
3. Aspects "revelatory of the phenomenon" are brought together.
4. A "situated structure" of the phenomenon is drawn up. This keeps close to the specific instance described by the research participant.
5. A "general description" of this is abstracted, shorn of the specifics which may be peculiar to the instance. Thus something like the essential features of such-and-such a psychological phenomenon is developed which can be regarded as the basis for future empirical study.

It is well worth considering Giorgi's process in conjunction with the very valuable paper of Wertz (1983), who details the attitudes activitives of the researcher in undertaking the analysis with sensitivity and skill. It is not a mechanical process of following the rules of analysis. Nor is it arbitrary.

van Manen (1990, 1991) is more literary in style, less prescriptive methodologically, and arrives not at essences but lengthy and extremely illuminating accounts of particular kinds of human situation (see Table 2.2).

Other contemporary phenomenological psychologists could probably be arranged, by methodological style, along a Giorgi–van Manen axis.

TABLE 2.2. Van Manen's method of interpretive writing (van Manen, 1990).

1. Formulate "phenomenological questions" (that is, questions relating to the nature of specific experiences) and bracket pre-existing assumptions about these experiences,
2. Engage in existential investigation
   - Generate data (in the same kind of ways as Giorgi)
   - Use the researcher's personal experiences
   - Trace etymological sources, believing, with Heidegger, that the meanings, especially the original meanings, of the words related to the phenomenon is revelatory unnoticed aspects of the experience.
   - Draw on literature, art and so on for similar reasons.
3. Write and rewrite—van Manen takes the view that, in the actual process of refining the careful expression of the experience, its meaning becomes clearer, or the associated meanings become more well-differentiated.

THE PRIMACY OF THE LIFEWORLD

Much contemporary phenomenological psychology, though close to Giorgi's approach does not in general seek essences or their equivalent. If general themes emerge from research, these may well be valuable. But to pre-suppose them is regarded as an avoidance of the epochē. As an initial move, interviews and the analysis of these are idiographic: it may be that *this* person's experience of whatever it is we are investigating is quite different to that of others. Work as skeptical as this regarding psychological universals or generalities means that when general themes do emerge the findings can be embraced with great conviction.

Increasingly we (Ashworth, 2003a, 2003b) have seen that interviews and their analysis can be strengthened and enriched by the recognition that the lifeworld has certain parameters or "fragments"[1] that can be expected to show themselves without hazarding the epochē. Because the lifeworld is universally present, and because these fractions are inevitable structures of the lifeworld—part of its essence, if you will—their recognition within the research does not introduce arbitrary presuppositions. Regrettably the classic phenomenological and existentialist authors do not provide a detailed account of the phenomenology of the lifeworld, though we do have good pointers to the essential features of the lifeworld in their work. Many writers on phenomenological psychology (Dahlberg, Drew, & Nyström, 2001; Pollio, Henley, & Thompson, 1997; Spinelli, 1989; Valle & Halling, 1989; van den Berg, 1972) mention certain—never all—of these features of the lifeworld: spatiality, temporality, embodiment, sociality, mood, historicity, and freedom.

The nearest person to enumerating a complete set akin to these is Medard Boss (1979)—translating Heidegger from the philosophical realm to that of psychological and medical science—but he includes speculative elements like being-towards-death which seem to us to entail more deeply interpretive thinking than the phenomenological description of that which is apparent would allow.

It is also important to mention that Schutz (Schutz, 1962, 1964, 1975; Schutz & Luckmann, 1974) has a set of "structures of the lifeworld" which are largely pre-suppositions allowing shared typifications of the social world. His account, however, answers to a different issue; it is a form of phenomenological sociology.

The point we are making is that *any* study of the lifeworld can be enriched by analysis in terms of these fragments. Indeed, they can be a means to the elucidation of the lifeworld. We contend that the correct approach to "something" as a feature of the *lifeworld* means that we address the phenomenon as a variant of the eidos "lifeworld." Lifeworld has essential features and is a human universal, and it is through the evocation of this structure that a particular empirical lifeworld can be described.

The wonder of Merleau-Ponty's writing—and this is especially true of the *Phenomenology of Perception* (1945/1962)—is that it cannot be read as meaningful unless the reader is in the phenomenological attitude themselves. It is not so much that he tells us directly how to see phenomenologically, but he leads us into the attitude by the places to which he guides our attention. So in the following exposition, I allow quotations from *Phenomenology of Perception* to sketch some of the meanings and issues surrounding the fractions of the lifeworld.

(a) Selfhood: What does the situation mean for social identity; the person's sense of agency, and their feeling of her own presence and voice in the situation? (For example, powerlessness might be a feature of the psychological situation for the individual.)

Identity is undeniably part of sociality—our identity links us to others and is provided by interaction with others:

> The body is no more than element of the system of the subject and his world, and the task to be performed elicits the necessary movements from him by a sort of remote attraction, as the phenomenal forces at work in my visual field elicit from me, without any calculation on my part, the motor reactions which establish the most effective balance between them, or as the conventions of the social group, or our set of listeners, immediately elicit from us the words, attitudes and tone which are fitting. Not that we are trying to conceal our thoughts or to please others, but because we are literally what others think of us and what our world is. (Merleau-Ponty, 1945/1962, p. 106)

But there is, with social selfhood, that awareness which poses ourselves as a *problem* co-extensive with the question of the meaning of the lifeworld.

> The central phenomenon, at the root of both my subjectivity and my transcendence towards others, consists in my being given to myself. *I am given*, that is, I find myself already situated and involved in a physical and social world... The fundamental power which I enjoy of being the subject of all my experiences, is not distinct from my insertion into the world. (Merleau-Ponty, 1945/1962, p. 360)

> [S]ince the lived is... never entirely comprehensible, what I understand never quite tallies with my living experience, in short, I am never quite at one with myself. (Merleau-Ponty, 1945/1962, p. 347)

(b) Sociality: How does the situation affect relations with others? There is no doubt of the intrinsic relatedness of one and the other:

> [I]t is precisely my body which perceives the body of another person, and discovers in that other body a miraculous prolongation of my own intentions, a familiar way of dealing with the world.... All of which makes another living being but not yet another man. But this alien life, like mine with which it is in communication, is an open life. It is not entirely accounted for by a certain number of biological or sensory functions.... There is one cultural object which is destined to play a crucial role in the perception of other people: language. In the experience of dialogue, there is constituted between the other person and myself a common ground; my thought and his are interwoven into a single fabric. (Merleau-Ponty, 1945/1962, p. 354)

(c) Embodiment: How does the situation relate to feelings about their own body, including gender, "disabilities" and emotions?

> The body is the vehicle of being in the world, and having a body is, for a living creature, to be involved in a definite environment, to identify oneself with certain projects, and to be continually committed to them. (Merleau-Ponty, 1945/1962, p. 82)

> To have a body is to possess a universal setting, a schema of all types of perceptual unfolding and of all those inter-sensory correspondences which lie beyond the segment of the world which we are actually perceiving. A thing is, therefore, not actually *given* in perception, it is to be internally taken up by us, reconstituted and experienced by us in so far as it is bound up with a world, the basic structures of which we carry with us, and of which it is merely one of many possible concrete forms. (Merleau-Ponty, 1945/1962, p. 326)

(d) Temporality: How is the sense of time, duration, biography apparent?

> [E]ach present reasserts the presence of the whole past which it supplants, and anticipates that of all that is to come, and by definition the present is not shut up within itself, but transcends itself towards a future and a past. (Merleau-Ponty, 1945/1962, p. 420)

(e) Spatiality: How is their picture of the geography of the places one needs to go to and act within, seen in the situation?

> [O]ne's own body is the third term, always tacitly understood, in the figure-background structure, and every figure stands out against the double horizon of external and bodily space. (Merleau-Ponty, 1945/1962, p. 101)

> Traditional psychology has no concept to cover these varieties of concept of place because consciousness of place is always, for such psychology, a positional consciousness [which is either a 'correct' or 'incorrect' representation] . . . Now here, on the other hand, we have to create the concepts necessary to convey the fact that bodily space may [for example] be given to me in an intention to take hold without being given as an intention to know. (Merleau-Ponty, 1945/1962, p. 104)

(f) Project: How does the situation relate to their ability to carry out the activities one is committed to and which are central to one's life? (Regrets? Pride?)

> [T]here is in human existence a principle of indeterminacy, and this indeterminacy . . . does not stem from some imperfection of our knowledge [regarding] . . . what we owe to nature and what to freedom. Existence is indeterminate in itself, by reason of its fundamental structure, and in so far as it is the very process whereby the hitherto meaningless takes on meaning, whereby . . . chance is transformed into reason; in so far as it is the act of taking up a *de facto* situation. We shall give the name transcendence to this act in which existence takes up, for its own purposes, and transforms such a situation. (Merleau-Ponty, 1945/1962, p. 169)

> The thing is inseparable from the person perceiving it, and can never be actually *in itself* because its articulations are those of our very existence, and because it stands at the other end of our gaze or at the terminus of a sensory exploration which invests it with humanity. (Merleau-Ponty, 1945/1962, p. 320)

(g) Discourse: What sort of terms—educational, social, commercial, ethical, etc.—are employed to describe—and thence to live—the situation?

> Speech is the surplus of our existence over natural being. But the act of expression constitutes a linguistic world and a cultural world, and allows that to fall back into being which was striving to outstrip it. (Merleau-Ponty, 1945/1962, p. 197)

Since Wittgenstein's "the limits of my language are the limits of my world" and his argument against the possibility of a private language, and since Heidegger's remarks about "language as the house of Being," and since Derrida's critique of Husserl as a so-called "philosophy of presence"—phenomenology is sometimes regarded as having been submerged in a move to discourse. But this is not the case. Phenomenological psychology retains emphatically the place of the conscious agent, and this agent is *intentionally* related to the world of experience, rather than a world of constructed discourse.

CONCLUSION

The authors I have discussed in these pages highlight important ways in which mainstream psychology from the start has set aside what many would regard as the key issue in the field, human meaningful experience. Early on, James pioneered psychological concern with experience through his development of the stream of consciousness and the self. Both these issues were developed by various schools of phenomenology, and in different ways by such authors as Allport, MacLeod, and Snygg and Combs with their various emphases on idiographic thinking and on the use of a "first-person" perspective.

Methodologically, the insistence that psychology should be concerned with experience in the sense of the description of things in their manner of appearing to awareness was carried through in great detail by the phenomenologists, following Husserl, and by the Gestalt psychologists. This is one emphasis that remains in qualitative psychology to this day. In a sense, this is a *perceptual* model of experience—the individual consciousness turned to some object of attention (and the prime feature of consciousness being intentionality—its relatedness to a world). A somewhat distinct strand of qualitative psychology, though there are authors who span the divide, emphasizes language or, more inclusively, discourse. Here we have a more obviously sociocentric view of human psychology, for which the model is less *perception* than *conception*. And the resources by which individuals can conceive the world, including themselves, are socially-available discursive practices.

Fundamentally, it appears to me, a psychology which looks to existential phenomenology for its philosophical inspiration will be qualitative (in focusing on meanings), will focus on the first-person perspective, will employ a form of epochē in order to set its sights firmly on these first-person meanings, and will be aware of the embeddedness of the phenomena under investigation within the individual lifeworld.

NOTE

1. We have called the elements of the lifeworld "fragments" to try—by this unfamiliar application of the term—to allow the reader to reflect on the fact that these are not independent categories or parameters, but are mutually entailed, with interpenetrating meanings.

REFERENCES

Allport, G. W. (1961). *Pattern and growth in personality*. London: Holt, Rinehart and Winston.

Allport, G. W. (1962). The general and the unique in psychological science. *Journal of Personality, 30*, 405–422.

Allport, G. W. (1965). *Letters from Jenny*. New York: Harcourt, Brace and World.

Ashworth, P. D. (1996). Presuppose nothing! The suspension of assumptions in phenomenological psychological methodology. *Journal of Phenomenological Psychology, 27*, 1–25.

Ashworth, P. D. (2003a). The phenomenology of the lifeworld and social psychology. *Social Psychological Review, 5*(1), 18–34.

Ashworth, P. D. (2003b). An approach to phenomenological psychology: the primacy of the lifeworld. *Journal of Phenomenological Psychology, 34*(2), 145–156.

Boring, E. G. (1950). *A history of experimental psychology* (2nd ed.). New York: Appleton-Century-Crofts.

Boss, M. (1979). *Existential foundations of medicine and psychology*. New York: Jason Aronson.

Brentano, F. (1995). *Psychology from an empirical standpoint*. London: Routledge. (Originally published 1874)

Bugental, J. F. T. (1964). The third force in psychology. *Journal of Humanistic Psychology, 4*, 19–25.

Bühler, C. (1971). Basic theoretical concepts of humanistic psychology. *American Psychologist, 26*, 378–386.

Combs, A. W., & Snygg, D. (1959). *Individual behavior. A perceptual approach to behavior* (2nd ed.). New York: Harper and Row.

Cumming, R. D. (1992). Role-playing: Sartre's transformation of Husserl's phenomenology. In C. Howells (Ed.), *The Cambridge companion to Sartre*. Cambridge: Cambridge University Press.

Dahlberg, K., Drew, N., & Nyström, M. (2001). *Reflective lifeworld research*. Lund: Studentlitterature.

de Beauvoir, S. (1988). *The second sex*. London: Picador. (Originally published 1949)

Fechner, G. T. (1966). *Elements of psychophysics*. New York: Holt, Rinehart and Winston. (Originally published 1860)

Giorgi, A. (1970). *Psychology as a human science: A phenomenologically based approach*. New York: Harper and Row.

Giorgi, A. (1985). Sketch of a psychological phenomenological method. In A. Giorgi (Ed.), *Phenomenology and psychological research*. Pittsburgh: Duquesne University Press.

Gurwitsch, A. (1964). *The field of consciousness*. Pittsburgh: Duquesne University Press.

Heidegger, M. (1962). *Being and time*. Oxford: Blackwell. (Originally published 1927)

Heider, F. (1958). *The psychology of interpersonal relations*. New York: Wiley.

Husserl, E. (1960). *Cartesian meditations: An introduction to phenomenology*. The Hague: Martinus Nijhoff. (Originally published 1931)

Husserl, E. (1970). *The crisis of european sciences and transcendental phenomenology*. Evanston, Ill.: Northwestern University Press. (Original partial publication 1936)

Husserl, E. (1977) Phenomenological psychology. The Hague: Martinus Nijhoff. (Originally published 1925)

Husserl, E. (1983). *Ideas pertaining to a pure phenomenology and a phenomenological philosophy* (First Book). Dordrecht: Kluwer. (Originally published 1913)

Husserl, E. (1989). *Ideas pertaining to a pure phenomenology and a phenomenological philosophy* (Second Book). Dordrecht: Kluwer. (Originally published 1952)

James, W. (1902). *The varieties of religious experience: A study in human nature. Being the Gifford lectures on natural religion.* London: Longmans, Green and Company.

James, W. (1950). *The principles of psychology* (Vol. 1). New York: Dover. (Originally published 1890)

Kvale, S., & Grenness, C. (1967). Skinner and Sartre: Towards a radical phenomenology of behavior? *Review of Existential Psychology and Psychiatry, 7*, 128–148.

Laing, R. D. (1965). *The divided self: An existential study in sanity and madness.* Harmondsworth: Penguin.

Lewin, K. (1935). *A dynamic theory of personality.* New York: McGraw-Hill.

Lewin, K. (1936). *Principles of topological psychology.* New York: McGraw-Hill.

MacLeod, R. (1947). The phenomenological approach to social psychology. *Psychological Review, 54*, 193–210.

Maslow, A. H. (1968). Toward a psychology of being (2nd ed.). Princeton: Van Nostrand.

Merleau-Ponty, M. (1962). *Phenomenology of perception.* London: Routledge and Kegan Paul. (Originally published 1945)

Miller, G. A., Gallanter, E. & Pribram, K. H. (1960). *Plans and the structure of behavior.* New York: Holt, Rinehart and Winston.

Misiak, H., & Sexton, V. S. (1973). *Phenomenological, existential and humanistic psychologies: A historical survey.* New York: Grune and Stratton.

New York Times. (1967) Obituary, Gordon Allport. [9th October]

Palmer, R. E. (1969). *Hermeneutics: Interpretation theory in Schleiermacher, Dilthey, Heidegger, and Gadamer.* Evanston, Ill.: Northwestern University Press.

Pollio, H. R., Henley, T., & Thompson, C. D. (1997). *The phenomenology of everyday life.* Cambridge: Cambridge University Press

Ricoeur, P. (1970). *Freud and philosophy: An essay on interpretation.* New Haven: Yale University Press.

Robinson G. D. (1995). Paul Ricoeur and the hermeneutics of suspicion: A brief overview and critique. Premise II, 8, 12 (electronic journal http://capo.org/premise/95/sep/p950812.html).

Rogers, C. R. (1967). *On becoming a peson: A therapist's view of psychotherapy.* London: Constable.

Sartre, J.-P. (1956). *Being and nothingness.* New York: Philosophical Library. (Originally published 1943)

Sartre, J.-P. (1957). The transcendence of the ego. New York: Farrar, Straus and Giroux. (Originally published 1937)

Schutz, A. (1962). *Collected papers, volume one: The problem of social reality.* The Hague: Martinus Nijhoff.

Schutz, A. (1964). *Collected papers, volume two: Studies in social theory.* The Hague: Martinus Nijhoff.

Schutz, A. (1975). *Collected papers, volume three: Studies in phenomenological philosophy.* The Hague: Martinus Nijhoff.

Schutz, A., & Luckmann, T. (1974). *The structures of the life-world.* London: Heinemann.

Skinner, B. F. (1964). *Behaviorism at fifty.* In T. W. Wann (Ed.), *Behaviorism and phenomenology.* Chicago: Chicago University Press.

Skinner, B. F. (1978). "Why I am not a cognitive psychologist." In *B. F. Skinner Reflections on Behaviorism and Society.* Englewood Cliffs, N.J.: Prentice-Hall.

Skinner, B. F. (1993). *About behaviorism.* London: Penguin.

Smith J. E. (1995). Freud, philosophy and interpretation. In L. E. Hahn (Ed.), *The Philosophy of Paul Ricoeur.* Chicago: Open Court.

Smith, J. A., Harré, R., & Langenhove, L. van (1995). *Rethinking psychology*. London: Sage.
Snygg, D., & Combs A. W. (1949). *Individual behavior*. New York: Harper and Row.
Spinelli, E. (1989). *The interpreted world: An introduction to phenomenological psychology*. London: Sage.
van den Berg, J. H. (1972). *A different existence: Principles of phenomenological psychopathology*. Pittsburgh: Duquesne University Press.
van Manen, M. (1990). *Researching lived experience*. New York: State University of New York Press.
van Manen, M. (1991). *The tact of teaching. The meaning of pedagogical thoughtfulness*. London, Ontario: The Althouse Press.
Valle, R. S., & Halliing, S. (1989). *Existential-phenomenological perspectives in psychology*. New York: Plenum Press.
Watson, J. B. (1913). Psychology as a behaviorist views it. *Psychological Review*, *20*, 158–177.
Wertz, F. (1983). From everyday to psychological description: Analysing the moments of a qualitative data analysis. *Journal of Phenomenological Psychology*, *14*, 197–241.
Wundt, W. (1904). *Principles of physiological psychology*. New York: Macmillan. (Originally published 1874)

CHAPTER 3

THE VALUE OF PHENOMENOLOGY FOR PSYCHOLOGY

AMEDEO GIORGI

INTRODUCTION

Phenomenology is a philosophy that had its beginnings in 1900 when Edmund Husserl (1900/1970), its founder, wrote one of his more famous works, *Logical Investigations*. The notion of phenomenology that Husserl advocated was more implicit in that work than defined, but his sense of phenomenology became explicit in his subsequent work, *Ideas I* (1913/1983). Phenomenology takes as its theme the relationship between conscious acts and the objects to which they are directed, the term object referring to anything, real or imaginary, or even illusory, that can be related to acts of consciousness. More briefly, one could say that phenomenology is the study of the relationship between consciousness and its objects. It is a philosophy that tries to develop this theme in a rigorous and consistent way, and it has also articulated an attitude and a method whereby the relationship between conscious acts and their objects can be accessed and thoroughly studied. Over the course of the last century phenomenology has developed in disparate and almost contrary ways, but as Spiegelberg (1982) has shown, there is nevertheless a unifying theme to its development.

Psychology also began as the study of mind, consciousness or experience—terms that are sprinkled throughout phenomenology as well. However, psychology's approach to consciousness was based upon one form or other of empiricism or naturalism—the prevailing outlooks of the Zeitgeist at the time. Psychology also adopted in an explicitly self-conscious way the natural scientific approach, which included in its academic setting, experimental situations, close and multiple observations, and a basic quantifying approach, although the latter was not exclusive in the beginning. It hoped to achieve the same success as the natural sciences did with nature by imitating as closely as possible the concepts, procedures and practices of the natural sciences. Physics, biology, and physiology were model sciences for psychology, when it began. However, what was not so clearly understood at the time was that in accepting the philosophical empiricism that drove the natural sciences, and in accepting its basic criteria, psychology was implicitly accepting physical objects as the model for psychical phenomena. And the parameters that were determinative of "physicality" were the parameters that came to determine how psychical phenomena were conceptualized. However, if there ever was a phenomenon that is not physical, it is consciousness (Husserl, 1913/1983; Sartre, 1943/1956). It is constantly in motion, it provides no sensory access to itself, and its specific causes are elusive. Yet, by means of assumptions and conceptualizations, psychology was able to study consciousness from an empirical perspective, although the price paid was severe distortion of the phenomenon.

Now, phenomenology is not anti-empirical, but it relativizes the empirical rather than making it the absolute basis of knowledge. We will make this clear momentarily, but first we want to indicate that while psychology began as the "science of consciousness or experience" (Wundt, 1901/1902), this definition has not been consistent throughout its recent history. Psychology is sometimes defined as the study of behavior (Watson, 1913) or even the unconscious (for many psychoanalysts). Our position is that phenomenology would be helpful whatever the chosen definition of psychology might be, although we limit ourselves in this chapter to a demonstration of phenomenology's helpfulness with respect to consciousness or experience alone. A consideration of behavior and the unconscious would make the chapter unduly lengthy. Phenomenology is a comprehensive approach so we believe that it could be helpful to psychology in a number of ways (methodologically, analytically, theoretically, descriptively, etc.) but space requirements limit us to the discussion of how phenomenological philosophy can be helpful with respect to the clarification psychological phenomena when consciousness or experience is its theme.

THE PHENOMENOLOGICAL APPROACH TO CONSCIOUSNESS

Since the phenomenological account of consciousness is the criterion by which we will judge the approaches associated with the founding of psychology, we will deal with it first. Our viewpoint is basically Husserlian.

However, certain cautions have to be specified before we begin. Firstly, Husserl was a consummate philosopher and everything he wrote was written with philosophical intentions in mind. While he made multiple references to psychology, at the beginning it was in terms of philosophical psychology—not the science of psychology, except some criticisms were addressed to the developing science—and later on the purpose of the references to psychology was to differentiate his mature sense of phenomenology from any type of psychology. His explicitly titled book *Phenomenological psychology* would also have to be considered a philosophical work (Husserl, 1962/1977). However, all that is written in this chapter on phenomenological psychology is from the perspective of the science of psychology, and more specifically, from the perspective of psychology as a human science. Thus, while inspired by him, we do not follow Husserlian theoretical points in a literal way because they are rarely geared toward scientific psychological issues.

Secondly, Husserl was breaking new ground with his philosophy and he often expressed matters ambiguously the first time around and he kept returning to certain areas of discourse so that he could clarify the issues more accurately, sometimes deepening the meaning of earlier concepts and at other times dropping certain terms and using newer ones. He described himself as an "eternal beginner." Consequently, nuanced differences sometimes surround Husserl's discussions of key points. All of these nuances could not be covered in a chapter of this size. What is true is that his thought was often original and groundbreaking, which is why contemporary thinkers still turn to him for guidance.

We shall indicate five features of consciousness that Husserl acknowledges are mostly missing from the writings of his psychological contemporaries. We believe that these are critical for a proper understanding of consciousness, and that psychology should embrace them. In general, Husserl (1970/1900, Vol. I, p. 535) distinguishes three meanings of consciousness, the most comprehensive referring to (1) the "entire real phenomenological being of the empirical ego, as the interweaving of psychic experiences in the unified stream of consciousness; (2) consciousness as the inner awareness of one's own psychic experiences; and (3) consciousness as a comprehensive designation for 'mental acts' or 'intentional experiences' of all sorts." Within that framework, there are five characteristics that we want to highlight.

Nonphysical Nature of Consciousness

Husserl makes a fundamental distinction between consciousness and objects that present themselves to consciousness. As Husserl (1913/1983, p. 89) himself expresses it,

> [T]here emerges a fundamentally essential difference between *being as a mental process and being as a physical thing.* Of essential necessity it belongs to the regional essence, Mental Process... that it can be perceived in an immanental perception; fundamentally and necessarily it belongs to the essence of a spatial physical thing that it cannot be so perceived. (Italics in original).

For what is ordinarily termed inner and outer perception Husserl uses the terms immanent, for the former, transcendent for the latter.

> [By] intentive mental processes related to something immanent, we understand those to which it is essential that their intentional objects, if they exist at all, belong to the same stream of mental processes to which they themselves belong (Husserl, 1913/1983, p. 79. Italics in original).

To say that intentive mental processes are directed to something transcendent is to say that the object being grasped does not itself belong to the same intentive process. Thus, "... all acts directed to essences or to intentive mental processes belonging to other Egos with other streams of mental processes, and likewise all acts directed to physical things or to realities of whatever sort..." (Husserl, 1913/1983, p. 79) would be transcendent. The basic point that we want to establish here is that the perception of a physical thing does not actually contain the physical thing because a mental process can only be combined with other mental processes within the same stream of consciousness. A physical object cannot become part of a stream of consciousness.

Having established the difference between immanent and transcendent objects Husserl goes on to note that the two types of objects present a fundamental difference in kinds of givenness. Physical things are perceived in an adumbrated way and mental processes are not perceived adumbratedly. To be given in adumbration means to be given in perspective, to be given partially and this implies that multiple perspectives on any spatial physical thing are possible. Multiple perspectives on a mental process are not possible because they are not given adumbratedly. That is, mental processes cannot present themselves as "something identical in modes of appearance..." (Husserl, 1913/1983, p. 96). Thus, Husserl calls the givenness of mental processes "absolute," meaning by that only that they cannot be given in sides or adumbratedly. Obviously, they can be accessed from different temporal perspectives.

Husserl also notes that neither physical things nor mental processes can be completely perceived, but for different reasons. A mental process is in flux and however good the reflective process is, it requires retention to keep passing phases present, and eventually, one has to pass on to retrospective recollection. Neither is the overall unity of our entire mental process graspable. With respect to transcendent physical objects, Husserl claims that we grasp the object itself (as opposed to some sign or image of it) but never completely. More perspectives giving other adumbrations are always possible. Consciousness and physical reality are not alike.

Consciousness Exceeds the Viewpoints of Naturalism and Empiricism

Husserl claims that a careful analysis of consciousness will show it to be richer and more complex than the perspectives of empiricism and naturalism will allow. Husserl gives credit to both of those philosophical movements for their intentions, but he notes that they have not freed themselves from all possible prejudices. Empiricism, of course is the philosophical movement that claims that the sole source of knowledge is experience and so knowledge is always *a posteriori*. Naturalism is defined as follows:

> Nature is thus conceived as self-contained and self-dependent... that the natural world is the whole of reality ... in thus restricting reality, naturalism means to assert that there is but one system or level of reality; that this system is the totality of objects and events in space and time; and that the behavior of this system is determined only by its own character and is reducible to a set of causal laws. (Runes, 1958, p. 205)

As stated, Husserl's point is that these two philosophical systems, however important they are for understanding physical nature, place constraints on the understanding of consciousness if the latter phenomenon is approached without prejudice and without prior commitments. To say that consciousness is natural in the sense of naturalism is to say that it is caused and that it presents no challenge to the system of natural events. But Husserl begins with exactly the opposite of this point: he contrasts consciousness to nature. He points out how differently they come to our awareness. Conscious processes can be known immanently, physical things cannot. The only way that one can have a memory of a specific tree is by having it become an immanent object in one's stream of consciousness. The actual transcendent tree which is the basis of the memory can be perceived by anyone close enough to it because it is in the world and not *in* anyone's consciousness.

For Husserl (1911/1965, p. 79) naturalism places primacy on "nature considered as a unity of spatiotemporal being subject to exact laws ... " While naturalism is correct when it deals with nature as defined above, it errs when it attempts to make its perspective universal and all-comprehending. Husserl (1911/1965, p. 79) makes this point well:

> [T]he natural scientist has the tendency to look upon everything as nature .... (and is) inclined to falsify the sense of what cannot be seen in (that) way.... Whatever is either itself physical, belonging to the unified totality of physical nature, or it is in fact psychical, but then merely as a variable dependent on the physical, at best a secondary 'parallel accompaniment.'

The consequence of this attitude is the "naturalizing of consciousness, including all intentionally immanent data of consciousness and... the naturalizing of ideas and consequently of all absolute ideals and norms." (Husserl, 1911/1965, p. 80). The primary difficulty is the falsification of sense or meaning that occurs. After all, Husserl (1911/1965, p. 90) reminds us, "Every type of object that is to be the object of a rational proposition, of a pre-scientific and then a scientific cognition, must manifest itself in knowledge, thus in consciousness itself, and it must permit being brought to givenness, in accord with the sense of all knowledge." Husserl's point here is that knowledge in the form of *consciousness of...* cannot be understood naturalistically. Husserl's arguments against naturalism cannot be repeated here in detail, but basically he shows that naturalism leads to a form of relativism which in turn leads to skepticism, which is untenable (see Husserl, 1900/1970 Vol. I, 1911/1965). We will now move to a critique of empiricism.

Within modern scientific thought it is generally accepted that the term "experience" is one of the widest concepts that can be used when speaking about a person's encounter with the world. However, that is not true of phenomenology. For Husserl, the main function of consciousness is to intuit, which in phenomenological parlance means "to make present." Consciousness is a presentifying medium. By means of consciousness we intuit, or are immediately present, to the world, to others and to ourselves. However, Husserl distinguishes types of consciousness, or intuitions, according to the object that is present. When a real or empirical object is given to consciousness, and by "real" Husserl (1913/1983) means any object that is in space, time and regulated by causality, he uses the term "experience" to refer to it. However, Husserl acknowledges that other types of presences also occur to consciousness and are capable of being intuited. As a philosopher he obviously found ideas ready-to-hand and so he also spoke of an ideative consciousness as a type of awareness that was different from experience because the given objects, ideas, are not real in the above sense. For one thing, ideas are not spatial. In brief, Husserl's point is that the range of objects that it is possible for consciousness to intuit is greater than merely real objects. In addition to ideas, one could point to memories, imaginary objects, anticipated events, meanings, and so on. These are all presences but they do not exist in the same sense as real objects do. Indeed, Husserl even goes so far as to speak of presences that are irreal—that is, they lack one of the defined characteristics of real objects. He (Husserl, 1913/1983, p. 35–36) writes:

> But what we tried to show in the preceding chapter was that by virtue of so-called eidetic seeing based on imaginings there spring from the imaginings new data, 'eidetic' data, objects that are irreal (*irreal*). But that, so the empiricist will conclude, is just 'ideological excess,' a 'reversion to Scholasticism' or to those 'speculative constructions a priori' in the first half [of] the nineteenth century by which an idealism, alienated from natural science, so greatly hampered genuine science.
>
> However, everything said here by the empiricist is based on misunderstandings and prejudices—no matter how well meant or how good the motive which originally guided him. The essential fault in empiricist argumentation consists of identifying or confusing the fundamental demand for a return to the 'things themselves' with the demand for legitimation of all cognition by experience. With his comprehensible naturalistic constriction of the limits bounding cognizable 'things,' the empiricist simply takes experience to be the only act that is presentive of things themselves.

Husserl's arguments lead to the position that neither naturalism nor empiricism are approaches that will allow a truly essential grasp of consciousness since it is neither physical nor naturalistic. Husserl argues for a phenomenological approach and a phenomenological approach necessitates the adoption of another attitude, the phenomenological attitude, which requires a departure from the comfort zone granted by the natural attitude (natural here understood as common or ordinary, not an attitude specifically geared toward "nature" in the above sense) which is the taken-for-granted attitude of everyday life and common sense. The assumption of the phenomenological attitude implies certain steps that help us overcome particular prejudices that have seeped into our everyday life attitudes. It calls for a certain fresh look at the world which means that one should bracket past knowledge about the phenomena of the world in order to be aware of present phenomena without the imposition of past experiences. It also implies the phenomenological reduction, which means first engaging with the phenomena and examining them without positing that they are real. To consider a "given" as a phenomenon means precisely to consider it as a presence rather than as something real or existing. By means of these methodological strategies Husserl argues that a more direct access to consciousness becomes possible because the strategies aim to distance consciousness as such (what he calls "pure" consciousness) from its intertwinings with empirical factors.

Consciousness Presents Objects of Many Types

Husserl claims that approaching consciousness from an empirical, naturalistic view gives privilege to one type of object—the real object that exists in space and time and is regulated by causality—and diminishes the characteristics of other types of objects. That is because one first looks at the characteristics of such an object and then one tries to ascertain how all of those characteristics make their appearance as knowledge for consciousness. But this procedure takes "existence" as the main feature of the object, whereas Husserl wants to make intuition, making present, the true function of consciousness, and the fact that the object that is present also exists, is a contingent factor, not at all a necessary one for the functioning of consciousness as such. Empiricism wants to account for the object as existing, as real, whereas phenomenology wants to account for its presence since existence is not a necessary factor for an object's presence to consciousness. An existing object considered as presence is what Husserl calls a phenomenon and that is why the phenomenological reduction is so important—it reduces existence to presence.

If it can be granted that the phenomenal realm is critical for understanding consciousness and the psyche, then Husserl makes a telling point about psychology's empirical approach. He (Husserl, 1911/1965, pp. 101–102) writes:

> The phenomenal had to elude psychology because of its naturalistic point of view as well as its zeal to imitate the natural sciences and see experimental procedures as the main point... [Psychology] has still neglected to pursue the question more profoundly, i.e., how, and by what method, can those concepts that enter essentially into psychological judgments be brought from the state of confusion to that of clarity and objective validity. It has neglected to consider to what extent the psychical, rather than being a presentation of nature, has an essence proper to itself to be rigorously and in full adequation investigated prior to any psycho-physics.

Husserl (1911/1965, p. 106) goes on to point out that a psychical phenomenon, as belonging to the phenomenal order, has "... no 'substantial' unity; it has no 'real properties,' it knows no real parts, no real changes and no causality; all these words are here understood in the sense proper to natural science." He (Husserl,

1911/1965, p. 107) then adds "It is the absurdity of naturalizing something whose essence excludes the kind of being that nature has." When examined as phenomena, the characteristics of various types of objects that can present themselves to consciousness become clearer. Some are irreal (ideas, fictive creatures); some are experiential objects only (dreams, images) and some are real (trees, tables). What would truly be psychological about each of these types of objects should have been psychology's first task, but it was bypassed.

CONSCIOUSNESS IS NOT GRASPED VIA APPEARANCES

Another difficulty in using physical things as the basis for understanding psychical phenomena is that they are given to consciousness differently than consciousness itself is given in reflection. We already noted above that Husserl used the mode of awareness as his evidence that consciousness was a type of being very different from real things. All things of the external world are given in adumbrations, in perspectives, so that what is given is usually a side, a partiality, and many perspectives are required to get a better sense of a transcendent thing. As Husserl (1913/1983, p. 97) says,

> It is indeed evident also that the adumbrative sensation-contents themselves, which really inherently belong to the mental process of perceiving a physical thing, function, more particularly, as adumbrations of something but are not themselves given in turn by adumbrations . . . while the perceived physical thing can be cancelled and regarded as non-existent, as an illusion, etc., the 'sensation-contents' themselves are beyond question in their absolute being.

By "absolute" here, Husserl means that sensations cannot be given adumbratively, one cannot take other perspectives on them. One has to accept them exactly as they present themselves and they do not come through appearances in Husserl's sense of the term, which means adumbratively. They are directly perceived and this is a function of consciousness that can never be captured by an empirical, naturalistic approach. Nor does our self-consciousness come through sensory givens. Consciousness knows itself by a type of direct intuition.

CONSCIOUSNESS IS INTENTIONAL

Another important feature of consciousness which is emphasized by Husserl but is skipped over by most of the schools of psychology is intentionality. By the intentionality of consciousness is meant the fact that consciousness is as a whole always directed to objects. Sometimes these objects are immanent, which means that the intentional object is part of the same stream of consciousness as the act, and sometimes the object is transcendent, which means that they are beyond consciousness. Still, the conscious act reaches the object itself and not some representative or image of the object. Intentionality is a way of describing the openness of consciousness to objects other than itself. It is by means of intentionality that conscious creatures are open to the world, others and even themselves.

Intentionality is a theme introduced to philosophical literature by Brentano, Husserl's teacher. However, Husserl modified Brentano's teachings regarding intentionality and developed the concept further. Husserl retains the idea that intentionality means directedness to an object, but he interpreted this relationship phenomenologically rather than naturalistically as Brentano had done. Mohanty (1972, pp. 55–56) has succinctly expressed the implications of this shift:

> For Husserl, the concept of intentionality demands a complete abandonment of the causal attitude in connection with conscious life . . . The mode of unity which prevails in one's conscious life may now be characterized by the adjective 'intentional'. Conscious states imply each other, lead to one another—not by mechanical association, nor by logical implication, but by motivating, anticipating and fulfilling each other. To understand a conscious state in this sense would be to follow all its *intentional implications.*
>
> States of consciousness are not outside each other and do not influence each other as physical states do. They rather internally refer to each other. It is for the psychologist to decipher such implications. (Italics in original).

The intentional feature of consciousness introduces a whole new mode of analysis.

Some Implications of Husserl's Analysis

What we have been doing in this section is highlighting dimensions of consciousness that a philosophical phenomenological analysis of that phenomenon would reveal. Before turning to the way the early psychologists treated consciousness, we will summarize the major points:

1. Consciousness is a type of being different from a physical real thing, so the latter cannot be a model for consciousness.
2. Consciousness, in opposition to empiricism, is not given to us via sensory experience.
3. Consciousness, in opposition to naturalism, is not given to us as a spatio-temporal entity fully regulated by causality.
4. The assumptions of naturalism result in a naturalizing of consciousness which violates its essence and retards a proper understanding of it.
5. Consciousness can present objects other than empirical ones, including irreal objects and experiential ones.
6. The immanent objects of consciousness are not presented adumbratively but directly.
7. The strategies involved with the assumption of the phenomenological attitude (bracketing, reduction) help attain a proper access to consciousness.
8. The phenomenological attitude reveals dimensions of consciousness hidden to empiricism and naturalism.
9. The phenomenal realm, accessed through the phenomenological reduction, is critical for understanding the functioning of psychological subjectivity.

10. The main function of consciousness is intuition, not experiencing, which is a mode of intuiting.
11. Appearances belong to external things that are given adumbratedly, but consciousness' awareness of itself is not through appearances.
12. Consciousness is intentional and the intentional object as such is not real.

A psychology that would theoretically allow the above features in its system would be much closer to the mark with respect to the phenomenon of consciousness. We wish to emphasize that while we have been critical of empiricism and naturalism it is not because they are wholly wrong. After all, consciousness does have some sort of dependency on nature the full sense of which still has to be worked out and consciousness does make contact with external, physical things. The criticism has been directed to the two philosophical systems as totalizing systems, as systems that require the reduction of consciousness to characteristics of the systems rather than trying to understand consciousness on its own terms.

PSYCHOLOGY'S APPROACH TO CONSCIOUSNESS IN THE FOUNDING YEARS

First of all, it should be made explicit that the framework within which most of the founding fathers approached psychological subject matter was that of the natural sciences. Wundt (1901/1902, pp. 22–25) of course came from a medical and physiological background and so he explicitly states that the natural sciences can serve as methodical models and he mentions physics and physiology as examples. Likewise, Titchener (1907, pp. 2–4) explicitly articulates those aspects of biology that it would be wise for psychology to imitate and Boring (1937, p. 473) has written that Titchener always used physics as a model science, Ebbinghaus (1908, p. 6) wrote that "natural science... served as a shining and fruitful example to psychology" and mentions physiology of the senses and the pathology of the nervous system as important exemplars. Yerkes (1911, pp. 25–26), in attempting to describe the goal of psychology, first describes the general aim of the natural sciences and then claims that psychology has the same general aim and he uses chemistry as an example. Finally, Woodworth (1921/1934, pp. 18–19) advocates the use of the experimental method for scientific psychology and goes to physics to describe a classical example. Of course, these psychologists represent the beginnings of natural scientific psychology and thus had to differentiate themselves from the earlier philosophical "arm-chair" systems, but the question is still open whether the turn to the natural sciences has offered better insights into psychological subject matter or merely different ones backed with the authority of the natural sciences. It is the opinion of this writer that the development of psychology as a natural science was more important for establishing the independence of psychology than for theoretical understanding of psychological phenomena. Obviously, the extent to which present day psychology still harbors the ambition to be a natural science and follows its developments and conceptualizations, then to that extent radical understanding of consciousness and subjectivity cannot happen. It further implies that psychology is still not truly an independent discipline that has clarified its own foundations.

We shall now consider how some of the early psychologists defined psychology in order to show how and why some of the features of consciousness highlighted by phenomenology cannot make themselves known. Obviously, we cannot cover these movements or schools thoroughly since that would require full chapters, or even books, on each movement. There is only space for the key points of each thinker from the perspective of consciousness.

The Approach to Consciousness by Structural Psychology[1]

Wilhelm Wundt

Wundt, of course, is regarded as the founder of modern experimental psychology, and he was an assiduous scholar whose approach to psychology was not narrow. Wundt noted that two definitions of psychology had been prominent up until his time but he believed that neither was satisfactory. The first definition he cites claimed that psychology was "... 'the science of mind,' psychical processes being regarded as phenomena from which it is possible to infer the nature of an underlying metaphysical mind-substance." The second definition he cited viewed psychology as "the science of 'inner experience,' psychical processes here looked upon as belonging to a specific form of experience... known through 'introspection'" (Wundt, 1901/1902, p. 1). Wundt disagreed with both definitions because the former belonged to philosophy and was metaphysically based and the latter implied that objects of the external world could not belong to psychological research, though Wundt believed that they could *whenever they were experienced.* For Wundt, the terms inner experience and outer experience did not refer to different kinds of objects but to different perspectives, different ways of looking at identical experienced objects.

While psychology was aligned with the natural sciences, Wundt distinguished the former from the latter by stating that the natural sciences concern themselves with the objects of experience considered as independent of the experiencing subject whereas psychology considers the objects of experience understood as being dependent on the experiencing subject. The perspective of natural science, he also called the perspective of mediate experience "since it is possible only after abstracting from the subjective factor present in all actual experience" (Wundt, 1901/1902, p. 3). Psychology's perspective he called that of immediate experience since it left the subjective factor intact. Wundt (1901/1902, p. 6) then said that "psychology is an empirical science which deals... with the immediate content of all experience" and it performs "empirical analysis and causal interpretation of psychical processes." Wundt (1901/1902, p. 10) believed that the psychology that he had initiated "could not admit any fundamental difference between the methods of psychology and those of natural science" and so he supported the use of experimentation along with introspection. While Wundt affirmed causal regulation of psychological processes, he distinguished between physical causality and psychological causality (Wundt, 1901/1902, pp. 357–368).

We can see from the above exposition that Wundt is thoroughly empirical and natural scientific. He discounted the definitions of psychology that preceded him and he believed that psychology was finally getting on to the right track by the empirical and scientific approach he advocated. He did acknowledge that psychology was different because of point of view, but not in terms of subject matter, and he explicitly rejected the claims of those psychologists who said that "inner experience" or "inner perception" was a special kind of reality. Husserl, of course, maintained that there was a difference in the way the two types of objects presented themselves, with the implication that there was some type of difference in *being*. Wundt's reasoning on this matter was in part due to his view that "outer objects" became psychological when they were experienced, i.e., somehow they then became real parts of consciousness. Thus, if the outer objects became psychological when they were taken up by consciousness, then the inner objects were no different because they, too, had to relate to the experiencing subjects as objects of consciousness. However, Wundt's "experienced objects as given to experiencing subjects" does allow for the possibility that non-empirical moments or constituents could also be discovered, but his empirical approach prevented him from seeing such possibilities.

In addition, Wundt's understanding of psychological subject matter as being "all of experience as dependent on the experiencing subject" could support the idea of intentional analysis since the idea of intentionality could hardly be better expressed: it is the object as experienced or as given to consciousness that matters. However, one would have to look for relations other than causality in order to discover the intentional relation. Wundt at least appreciated that the "causality" that governed psychological processes was different in principle from physical causality. Wundt believed that immediate experiences were compounds that were developed by a process of "creative synthesis" and that psychical measurement had to do with qualitative values. Husserl was freer in admitting that there were relations among humans, or in the world, that were other than causal ones.

Edward B. Titchener

Titchener was a student of Wundt who founded a lab at Cornell University and spent his entire career there. Next to Wundt, he was probably the most famous psychologist in the world, especially between 1900 and the mid 1920s. He also trained many students who went on to establish psychological labs in universities throughout America. He was highly influential in giving psychology an experimental direction and was a primary interpreter of Wundt to Americans, although, subsequently, what he said about Wundt has proven to be largely erroneous (Blumenthal, 1975, 1980; Danziger, 1979).

Like James, Titchener believed that psychology was the scientific study of mind, by which term he meant that the sum total of thoughts and feelings that any individual could have. He emphasized, however, that the mind did not "have" thoughts and feelings, but that it *was* "thoughts and feelings" (Titchener, 1907, p. 5). While mind is the integrative term for Titchener, he goes on differentiating mind into the child, adult and senile mind. Further, for Titchener (1907, pp. 19–20), "Each of these part-minds consists of a series of consciousnesses. By *consciousness* we mean 'mind now'; the mind of the present moment... The mind at every 'now,' whether it be in childhood or in manhood or in old age, is a consciousness" (Italics in original). Thus, concretely a psychologist only studies specific consciousnesses and every consciousness is then conceived by Titchener to be made up of a number of concrete processes, such as wishes, feelings, ideas, etc. Titchener then moves on to the next level of analysis and states that

> no concrete mental process, no idea or feeling that we actually experience as part of a consciousness, is a simple process, but that all alike are made up of a number of really simple processes blended together. (Titchener, 1907, p. 21).

Thus we have arrived at the famous mental elements, sensations, and affections, and the task of psychology was to build up mind by reversing the above analytic process.

The above description of the Titchenerian system, however brief, should make clear that Titchener takes "consciousness now" as a totality and then breaks it down to its most basic elements and then tries to reconstruct each "part-mind." This is undoubtedly the tried and true procedure of natural science, but does it do justice to the phenomenon of consciousness? We will respond to this question shortly below, but it is important to realize that Titchener was as interested in establishing psychology as a natural science on the basis of experimentation as he was interested in the definition of psychology. In his *Primer* of 1907, he gives as much space to the discussion of science as to psychology. He used as a criterion the science of biology and he did the same when he differentiated his system of psychology from functionalism (Titchener, 1898, 1907).

Now, one of the problems with Titchener's approach to psychology is that there is no account of intentionality. His psychology was known as a "psychology of content" rather than a functional one or a psychology of act. In part, this is due to Titchener's strong commitment to the natural science approach. He (Titchener, 1907, pp. 6–7) writes:

> 'The objects of which science treats are of two different kinds. They may be *things*, or they may be *processes*.... Things are... lasting and unchanging.... Processes are always changing.... Psychology is a science that treats exclusively of matters of the second sort, i.e., of processes' (Italics in original).

If consciousness was a process, then it had to be a process in the way that things generated or were related to processes. Thus consciousness was an analog of a thing: it was a self-enclosed process. A stimulus from the outer world could trigger it off (as well as inner stimuli) but then it remained a question of how to determine the basic elemental processes and to follow their development and integration into the complex totality that any "consciousness now" was. The activity of consciousness for Titchener was a process within the totality, but not an activity relating to the outside world with the initiative on the side of consciousness as intentionality would imply.

Titchener fought against any aspect of the competing systems of his time that he believed would compromise the scientific status of his own approach to psychology. He was naturalistic in outlook and empirical in the sense of this chapter, but he argued for an even stricter experimental approach, as represented by Wundt, rather than the older empirical, more philosophical psychology that Brentano represented (Titchener, 1921a). Titchener (1921b) was against functional psychology because he believed that it was empirical in Brentano's sense and because it emphasized the appearance of consciousness as an adaptive strategy of the organism in relation to the world rather than a phenomenon to be investigated in its own right. In addition, for Titchener, the adaptive attitude of functional psychology made it seem as though functionalism was flirting with teleology which Titchener was forcibly against since he thought that there was no place for teleology in science. Carr (1925), writing in response to Titchener's critique denied the teleological accusation, but his argument did not sway Titchener.

Titchener (1921b, 1929/1972) had several arguments against act psychology. His first complaint was that there was not so much an act psychology as act psychologies. He (Titchener, 1929/1972) started with Brentano and showed that how his school separated into Meinongian and Husserlian wings and then he examined the works of Stumpf, Lipps, Witasek, Messer, Kulpe, etc., and showed that there was no consistency among them. In general, in his view, act psychology was wholly dedicated to the activities of consciousness, which he interpreted as the description of the mental functions of the organism, which then became in effect a functional psychology even if it was not biologically based. Titchener did not seem to realize that "act" for these thinkers was more actualization than activity and an act's status was determined by regressing back from the object itself.

Titchener (1921b, pp. 370–371) quotes Brentano on intentionality and then writes:

> This is evidently the language of function, not of structure. Indeed, Brentano uses the phrases *psychisches Phanomenon* and *Seelenthatigkeit* interchangeably; his 'fundamental' or 'principal classes of psychical phenomena' are the 'mental activities' of ideation (not 'idea!'), judgment and interest. The spirit of his whole psychology is physiological.... Now the mental elements of the experimentalists, the bare sensation and the bare feeling, are abstractions, innocent of any sort of objective reference.

Thus, Titchener explicitly wants to state that his elements, or elemental processes, do not have reference to objects, the cornerstone of the approach of act psychology when it acknowledges intentionality.

Titchener's entire system is constructed in this fashion. While one may easily understand how sensations and affections may be analyzable without references to intentionality, Titchener is also able to account for the phenomena of attention and meaning, phenomena that one would think should relate to the world, without actually referring to intentionality. The former is reduced to the attribute of clearness and the latter is accounted for entirely in terms of context (Heidbreder, 1933). Titchener also disagreed with the Wurzburg school of imageless thought because he did not want to add a "thought element" to sensations, images and affections that were impalpable.

Perhaps the most obvious illustration of Titchener's system is his discussion of the stimulus error. There is a long history to the stimulus error (Boring, 1921) but Heidbreder (1933, p. 129) presents the issue straightforwardly and clearly:

> When a person observes naively, in the manner of common sense, he sees, for example, a table; but if an introspectionist sees a table while making a scientific observation of his perception of a table, he is committing the stimulus error. He is attending to the stimulus instead of the conscious process the stimulus occasions in him; he is reading into the process what he knows about the object and, in doing so is failing to distinguish what he, as a person, knows about the stimulus from what he, as an observer, is immediately aware of in his experience. All that his immediate experience gives him is color and brightness and spatial pattern. The rest is interpretation, not observation, and is the result of reading meaning into what is immediately present in his experience.... [It is] to see the thing, rather than the conscious content...

So, while naïve consciousness gives us the world of everyday life, Titchener wants the psychologist to withdraw from such awareness and to concentrate only upon what is "really" given, internal to consciousness, and how it is given there. This is the psychologist's special contribution via the method of introspection. Obviously, in this view, the world gets represented in the mind.

It should be clear that Titchener's strong commitment to the experimental method and the natural scientific approach limited his discoveries about consciousness to what the method and attitude allowed. He looked for what he termed the existential elemental processes and wanted to discover the laws guiding their development to the complex totalities that are given in everyday experience. He wanted to abstract from the everyday meaningfulness of objects in order to describe the existing elemental sensory attributes that helped constitute the experience. He would not allow anything into his system that would challenge his natural scientific outlook.

Husserl, also confronted with a consciousness that was ever moving and seemed to be elusive, approached the problem differently. He (Husserl, 1950/1960, p. 49) writes:

> ... [S]ince the realm of phenomena of consciousness is so truly the realm of a Heraclitean flux. It would in fact be hopeless to attempt to proceed here with such methods of concept and judgment formation as are standard in the Objective sciences. The attempt to determine a process of consciousness as an identical object, on the basis

> of experience, in the same fashion as a natural Object—ultimately then with the ideal presumption of a possible explication into identical elements, which might be apprehended by means of fixed concepts—would indeed be folly. Processes of consciousness—not merely owing to our imperfect ability to know objects of that kind, but a priori—have no ultimate elements and relationships . . . in their case, it would be rational to set ourselves the task of an approximative determination guided by fixed concepts . . . the idea of an intentional analysis is legitimate, since, in the flux of intentional synthesis . . . *an essentially necessary conformity to type* prevails and can be *apprehended in strict concepts.* (Italics in original.)

Thus, Husserl looked for concepts where Titchener was seeking substantial or really existing stuff. Husserl also, as we saw, allowed for irreal objects and intentional objects that were not reducible to palpable contents. The very idea that consciousness could present phenomena that were other than real was anathema to Titchener.

The ultimate irony for Titchener's type of psychology is that the program of research failed and terminated. Hardly anyone followed Wundt's experimental voluntarism, the Wurzburg school could not get past its controversy about imageless content and Cornell structural psychology died with Titchener in the late 1920s. Recall that Titchener had criticized act psychology because there was little agreement among authors and the theories were too system dependent. Titchener argued that he could appeal to the experimental facts of his structural psychology. But one of the reasons that introspection failed was that the different researchers from different labs, and therefore different perspectives, could not agree on the facts—not even on the number of basic elements. Titchener did not realize that the experimental facts also depended upon the systemic view that the researcher held. It was the fundamental outlook shared by all early investigators that was flawed.

THE APPROACH TO CONSCIOUSNESS OF THE FUNCTIONAL SCHOOL

The school of functionalism—initially centered at the University of Chicago with the work of John Dewey, James Angell and later, Harvey Carr—was a reluctant school in the sense that it came into being as a response to Titchener's (1898) critique of psychological practices that he considered to be at deviance from his understanding of scientific psychology. But perhaps functionalism was simply an American interpretation of psychology since it fitted in much better with the American pragmatic character.

Angell's (1907) response to Titchener's critique made three basic points: (1) functional psychology wanted to portray "the typical operations of consciousness under actual life conditions"; (2) functional psychology was part of the growing interest in evolutionary hypotheses, and so it viewed organic structures and functions as possessing characteristics that survived the evolutionary process and fitted in with the current environment, and (3) functionalism had a special interest in mind–body problems. This school noted that consciousness seemed to appear when there were problems in the organism's relationship to the environment and that it tended to disappear as actions became automatic. Thus, consciousness was seen as adaptive or adjustive, but the main interest was in the activity of the organism in relation to its environment.

Functionalism, following Darwinian leads, was basically naturalistic and saw no difference in the appearance of consciousness than with any other evolutionary mutation that might have emerged. Indeed, as Heidbreder (1933, pp. 217–218) points out,

> ... functionalism does away with the dualism which asserts that the physical and the mental are two different orders of events. It regards the distinction between mind and body as a convenience in our thinking ... but not as one that should prejudice us toward the belief that mind and matter are two really different entities.

Basically, the functionalists treat the distinction between mind and body as methodological since both can be adaptive to situations in the world, and that was where the functional interest lay. Its perspective was large and so it sought to understand how consciousness functioned within the larger context of person–environment relationships. It was not interested in probing the characteristics of consciousness or in trying to determine what was essential about it. It was an almost wholly biological perspective.

Nevertheless, functionalism's perspective could support some of the features of psychological research that phenomenology values. The very fact that they show interest in practical issues and in organism–environment relationships implicitly shows a Lifeworld orientation. Titchener and Wundt were biased toward a laboratory situation, but the functionalists, while not being against experimentation, did not emphasize experimentation almost to the exclusion of everything else as Titchener did or Wundt did more moderately. In addition, the subject–world relations, they emphasized could support an intentional relationship and the body–mind relationships they spoke about from an "identity perspective" could have benefited from phenomenology's notion of a layered subjectivity. However, their naturalistic, biological, and empirical orientation prevented the emergence of phenomenological insights.

JAMES'S APPROACH TO THE PHENOMENON OF CONSCIOUSNESS

James (1890/1950) is known to be an unsystematic writer, so much so that he is often taken to be a precursor for many subsequent movements in psychology, from behaviorism to phenomenology. Undoubtedly there is some truth in these retrospective evaluations for, as Allport (1943) has pointed out, James's writings contained many "productive paradoxes." However, James was fairly clear about many of the characteristics of consciousness.

James (1890/1950, p. 1) like many of his contemporaries , defined psychology as the "Science of Mental Life, both of its phenomena and their conditions." The phenomena for him were experiential activities like feelings, desirings, decision makings, etc., and the conditions were bodily processes, especially those in the brain. James often spoke about "thinking and thoughts" when he clearly meant consciousness or experience (Linschoten, 1968, pp. 40–67) for he was looking for a generic term to refer to all psychical phenomena. As is well-known, he used the metaphor of "stream" to account for how consciousness was always in flux and he described substantive and transitive parts of consciousness as equally important. He then attributed five characteristics to the stream of consciousness (James, 1890/1950, p. 225):

1. Every thought tends to be part of a personal consciousness.
2. Within each personal consciousness thought is always changing
3. Within each personal consciousness thought is sensibly continuous.
4. It always appears to deal with objects independent of itself.
5. It is interested in some parts of these objects to the exclusion of others, and welcomes or rejects—*chooses* from among them, in a word—all the while. (Italics in original.)

Later as James moved to his position of "radical empiricism," he tried to move away from any type of dualistic position. He came to explicitly deny that any type of mental substance existed and he tried to account for the relationship between the "knower and known" in terms of the same sort of experiential "stuff." In James's (1912/1958, p. 37) own words:

> [B]ut breath, which was ever the original of 'spirit,' breath moving outwards, between the glottis and the nostrils, is, I am persuaded, the essence out of which philosophers have constructed the entity known to them as consciousness. *That entity is fictitious, while thoughts in the concrete are fully real. But thoughts in the concrete are made of the same stuff as things are.* (Italics in original.)

James tries to explain the meaning of his conclusion when he (1912/1958, p. 3) writes:

> Let me then immediately explain that I mean only to deny that the word stands for an entity, but to insist most emphatically that it does stand for a function. There is, I mean, no aboriginal stuff or quality of being, contrasted with that of which material objects are made: but there is a function in experience which thoughts perform, and for the performance of which this quality of being is invoked. That function is *knowing.* (Italics in original.)

James posits only one type of "stuff" of which the world is made and he calls that "stuff" pure experience and that is his way of solving the problem of consciousness and knowledge. Nothing more needs to be added.

It can be seen from the above that James's perspective on consciousness is wholly empirical since it is experience that survives his deletions. But since his approach is also descriptive, he does bring some different dimensions of consciousness to the fore that the structuralists overlooked. He emphasized that consciousness was personal and that it is selective. He reiterates as well the facts that consciousness is constantly changing, is continuous and that it relates to objects that are independent of it. For James, consciousness's awareness of itself comes through introspection. None of these features are foreign to a phenomenological approach. Husserl (1962/1977, 1952/1989) has written about the personal nature of consciousness and the fact of a personal ego, and the idea of selectivity is implicit with the idea of adumbration and the distinction between theme and horizon. Also, Husserl's account of consciousness's ability to be aware of itself would be in terms of reflection and reflexivity.

James (1890/1950, pp. 183–194) also devoted some space to the fundamentals of scientific psychology. The minimum components that he considered to be necessary in order to have a viable psychology were four: the psychologist; the thought (consciousness) studied; the object of the thought (consciousness) being studied, and finally what he called the psychologist's reality. Recently, a phenomenological perspective was brought to this schema (Giorgi, 1992) and as a result of the introduction of the notion of intentionality, the schema was modified. Phenomenologically modified, the James's schema would be as follows. As many of the founders believed, the psychological perspective for James depended on point of view, and that point of view is what constitutes the psychologist's reality. The psychologist, whether studying himself or another, deals with how the act of consciousness relates to the object that is presented to the experiencer from the viewpoint of subjectively constituted meanings by the experiencer. James wanted to account for the act–object relationship in terms of the knowing relationship, and even though he was aware of the work of Brentano and Husserl, he obviously was not motivated to work through the question of the intentional relationship even though his schema is set up to do so. Knowledge depends upon the intentional relationship, but it is too narrow to carry the project of psychology. Thus, James's descriptive approach introduces many themes sympathetic to phenomenology, but in our opinion, he fell short of being able to articulate the maximum possibilities because of his radical empiricism. He even acknowledged that from the perspective of empiricism consciousness as such appeared irreal, but he ended up calling it fictitious as a whole despite his recognition of individual thoughts.

THE APPROACH TO CONSCIOUSNESS ON THE PART OF ACT PSYCHOLOGY

Act psychology is the term given to the school initiated by Franz Brentano. Titchener called it empirical psychology as opposed to experimental psychology and he believed that the force of the school was due to its use of applied logic rather than the accumulation of experimental facts. Ultimately Titchener thought that the school failed because it only indulged in system building without a basis of "independent facts," but we saw that all of Titchener's facts did not prevent the demise of his own perspective.

Phenomenology itself grew out of this tradition for Husserl, too, was a student of Brentano. Brentano's students were mostly philosophers, but there was a more direct, but still slight, influence of Husserl on psychology, although how correctly psychologists understood the phenomenological project is questionable. The reason is that Husserl wrote primarily for philosophers. In order for psychologists to apply phenomenological concepts, elaborate interpretations of Husserlian themes were required. The other reason for doubting a strong influence was that the Wurzburg school, the school of psychology we are discussing, flourished from about 1900 until about the beginning of World War I, and Husserl's more explicitly phenomenological work, *Ideas I,* did not come out until 1913. The work of Selz, a follower of Meinong, was undertaken in the 1920s but his work was only belatedly appreciated after World War II (Humphrey, 1963). This means that the Wurzburg school was waning while phenomenology was waxing, so the exchange could not have been too rich. The phenomenological ideas that the psychologists had to work with were more suggestive than explicit. Perhaps that is why the Wurzburg school remained basically empirical even though they were on the threshold of eidetic discoveries.

The work of the Wurzburg school is covered exceptionally thoroughly by Humphrey (1963). The psychology department at the University of Wurzburg was chaired by Kulpe, who was trained by Wundt and G. E. Müller, but he took an interest in the processes of thinking and these studies led him from Wundt's perspective on psychology to one that was closer to Brentano (Boring, 1929/1950, p. 409). Humphrey (1963, pp. 30–31) reviews how during the first decade of the 20th century some half dozen studies on thinking were conducted at the psychological laboratory at Wurzburg, in order to understand how the laws of association worked with the phenomenon of thinking. The outcome was that the association theory of thinking was first challenged, and then overturned, and its basic elements—sensations and images—were not deemed sufficient to account for the introspections given during the process of conducting certain thoughtful tasks. The subjects participating in those tasks described the presence of certain "dispositions of consciousness" (*Bewussteinslage,* as translated by Mandler & Mandler, 1963) that were given in their experience of the task and yet were not reducible to sensory contents or images. Later, these dispositions were described as simple "awareness" and even as certain "determining tendencies" required for the completion of the tasks. However, contrary to prevailing theory, they were nonsensory and not further analyzable and sometimes they were treated as being unconsciously inferred, i.e., as a priori to the task, but necessary for the completion of it. Titchener described these findings as "impalpable (his term for *unanschaulich),* but phenomenologists would translate the term *unanschaulich* as nonintuitable. Nonintuitable, that is, for those seeking empirical givens, but not necessarily so from a phenomenological perspective that acknowledges intuitable irreal objects. The dominance of the empirical perspective here was another case of theory triumphing over findings.

Selz's work, which occurred after Kulpe left Wurzburg, broke completely with associationism and converged with the work of the Gestalt psychologists, and because the impact of the latter school was greater, the focus shifted to the Gestaltist's efforts concerning the process of thinking (Mandler & Mandler, 1963). The Wurzburg school is often cited as one of the reasons that the introspective method failed because it seemed that what was important for thinking to take place was primarily unconscious (the "set" that prepared one for the task; the determining tendency; the rules governing the process, etc.) and all introspection could do was provide the facts of which one was aware. Actually, Husserl (1911/1965, p. 115) had warned scholars not to confuse phenomenological intuition with introspection because the latter "looks only at the conscious activity, not at its objective correlate (noema)." Moreover, in order to overcome all of the idiosyncratic variations of individual subjects one would have to grasp the essence of the thinking processes and not remain only with the facts. Clearly, had the Wurzburgers utilized a phenomenological approach, they would have come up with very different interpretations of their findings.

As it is, it is important to remember that the series of experiments performed by the Wurzburgers were unorthodox in several ways. First of all, the motivation was to understand the processes of association during thinking from a qualitative perspective. Had they simply used quantitative analyses of association, most probably there would have been no controversy. It was the descriptions of what was going on during thinking that was the basis of the controversy. The Wurzburg psychologists began as associationists and yet as Mayer (1983) points out, their results were anti-atomistic, anti-mechanistic, anti-empirical, and anti-imagery. All of these perspectives were contrary to associationism as understood at the time. Wundt also criticized them for not being good experimenters and for failing to use the introspective method correctly (the Wurzburg psychologists modified introspection so that it became systematic controlled introspection with a different goal from the Wundtian interpretation). However, Humphrey (1963, pp. 106–119) makes clear that Wundt exaggerated the so-called defects and very little remains of his criticisms when carefully examined. Humphrey demonstrates that Wundt's criticisms made in 1907 touched upon the introspective method *as such* and not just on how the Wurzburgers used it. Watson (1913) made the same criticisms against the method of introspection a half dozen years later when he initiated Behaviorism. When one adds that the Wurzburg psychologists also wanted to affirm intuitable data that were not palpable (i.e., empirical) one can see why they ran into trouble. They were challenging many of the sacred tenets of the received wisdom concerning the practices and results of scientific psychological work. Titchener, too, criticized their work, trying to either fault the descriptions of the Wurzburg subjects or else saying that the results were due to kinaestheses that the subjects missed. Titchener could not allow "impalpable" data. In any case, the so-called "failure" of the school may be just as much due to its heterodoxy as to its use of introspection because since the ascent of the cognitive perspective in psychology, greater appreciation of Wurzburg achievements is in evidence.

Kulpe (1912/1964, p. 213) seemed to be aware of the difficulties involved when breaking new ground in scientific contexts, but he did not know exactly how to give voice to the new consequences:

> [The psychology of] thinking unlocked the door to the true internal world, and it was no mysticism that led us there, but the abandoning of a prejudice. Bacon already knew that the road to truth is paved with prejudices. In the present instance they happen to derive from the exact natural sciences, for whom in the last decades sensory observation meant everything and for whom concepts were only an expedient used to represent, in the simplest possible fashion, facts based on sensory experience.

Kulpe was aware of the limits he was encountering, but it seems that he himself did not know how to go beyond them, except negatively. He was aware of Husserl's work through Buhler, but did not seem to work it through himself (Boring, 1929/1950). Humphrey (1963, p. 77) explicitly states that the significance of Husserl's work for their research seemed to have bypassed the Wurzburgers.

Obviously, a phenomenological approach would have been far more sympathetic for the discoveries of the Wurzburg school. For phenomenology, intuition is not limited to empirical data and irreal objects could be acknowledged. In addition, the reflexivity of consciousness (not reflectivity) could also have been brought into play. Reflexivity refers to that characteristic of consciousness whereby we often know exactly what we are doing prior to all reflection. Finally, the introspective method tried to garner all the facts and then stopped there. Phenomenology interrogates those facts, in order to come up with their essence, another irreal given, obviously dismissed by scientists of that era.

CONCLUSION

To the extent that consciousness or experience is to be part of the subject matter of psychology, then it behooves psychologists to have a proper understanding of such phenomena. That consciousness phenomena are more difficult to apprehend and do not appear statically is acknowledged by most thinkers, but that fact has not motivated such thinkers to give up the accepted constraining framework: realism, empiricism, naturalism, and the natural scientific model. Those perspectives are desirable when dealing with things and nature, but they impose constraints when trying to deal with presences that are not real. Phenomenology is a rigorous philosophy that tries to expand the philosophical outlooks so that all possible experienceable or intuitable phenomena can receive rigorous treatment. Dealing with phenomena, i.e., objects as they are strictly experienced, is one way to overcome the realistic bias.

Our brief survey indicated that Wundt and Titchener described empirical realities in such a way that the intentional relationship, the intentional object and careful description of the status of phenomena could have emerged, but they did not; the functionalists Angell and Carr had the possibility of announcing the role of the lifeworld and the intentional relationship, but they did not; James came close to recognizing that consciousness could harbor irreal dimensions and that such irrealities could be intuited, but he fell short; finally, the Wurzburg school encountered irreal objects and the horizonal aspects of conscious activity, and at times explicitly referred to intentional objects, but could move no further. Our claim is that these features of consciousness cannot be ignored if a correct understanding of consciousness or experience is to be achieved and a phenomenological perspective will permit these achievements of consciousness to be properly understood and utilized.

NOTE

1. While the term "structural psychology" clearly fits Titchener's program since he himself coined the term, it is somewhat doubtful in the case of Wundt. However, Titchener did include Wundt in his perspective, however erroneously, and so we let the term stand for the two of them. Obviously, the sense of structural used here is quite different from the structuralism that developed later in the century in anthropology and Gestalt theory, etc.

REFERENCES

Allport, G. (1943). The productive paradoxes of William James. *Psychological Review, 50*, 95–120.

Angell, J. R. (1907). The province of functional psychology. *Psychological Review, 14*, 61–91.

Blumenthal, A. L. (1975). A reappraisal of Wilhelm Wundt. *American Psychologist, 30*, 1081–1088.

Blumenthal, A. L. (1980). Wilhelm Wundt and early American psychology: A clash of cultures. In R. W. Rieber & K. Salzinger (Eds.), *Psychology: Theoretical-historical perspectives.* New York: Academic Press.

Boring, E. G. (1921). The stimulus error. *American Journal of Psychology, 33*, 449–471.

Boring, E. G. (1937). Titchener and the existential. *American Journal of Psychology, 50*, 470–483.

Boring, E. G. (1950).*A history of experimental psychology.* New York: Appleton-Century-Crofts. (First edition work published 1929)

Carr, H. (1925). *Psychology: The study of mental life.* New York: Longmans, Green and Co.

Danziger, K. (1979). The positivist repudiation of Wundt. *Journal of the History of the Behavioral Sciences, 13*, 205–230.

Ebbinghaus, H. (1908). *Psychology: An elementary text-book* (M. Meyer, Trans.). New York: D.C. Heath and Co.

Giorgi, A. (1992). A phenomenological reinterpretation of the Jamesian schema for psychology I. In M. E. Donnelly (Ed.), *Reinterpreting the legacy of William James* (pp. 119–136). Washington, D.C.: American Psychological Association.

Heidbreder, E. (1933). *Seven psychologies.* Englewood Cliffs, NJ: Prentice-Hall.

Humphrey, G. (1963). *Thinking: An introduction to its experimental psychology.* New York: Wiley.

Husserl, E. (1960). *Cartesian meditations* (D. Cairns, Trans.). The Hague: Nijhoff. (Original work published 1950)

Husserl, E. (1965). Philosophy as a rigorous science. In Q. Lauer (Ed. & Trans.), *Phenomenology and the crisis of philosophy* (pp. 71–147). New York: Harper Torchbooks. (Original work published 1911)

Husserl, E. (1970). *Logical investigations Vols. I and II.* (J. N. Findlay, Trans.). New York: Humanities Press. (Original work published 1900)

Husserl, E. (1977). *Phenomenological psychology* (J. Scanlon, Trans.). The Hague: Nijhoff. (Original work published 1962)

Husserl, E. (1983). *Ideas pertaining to a pure phenomenology and to a phenomenological philosophy. First Book* (F. Kersten, Trans.). The Hague: Nijhoff. (Original work published 1913)
Husserl, E. (1989). *Ideas pertaining to a pure phenomenology and to a phenomenological philosophy. Second Book* (R. Rojcewicz & A. Schuwer, Trans.). Dordrecht: Kluwer Academic Publishers. (Original work published 1952)
James, W. (1950). *Principles of psychology*. New York: H. Holt and Co. (Original work published 1890)
James, W. (1958).Does consciousness exist? In W. James, *Essays in radical empiricism.* New York: Longmans, Green & Co. (German original published 1912)
Kulpe, O. (1964). The modern psychology of thinking. In J. M. Mandler & G. Mandler (Eds.), *Thinking: From association to Gestalt*. New York: Wiley. (Original work published 1912)
Linschoten, H. (1968). *On the way toward a phenomenological psychology: The psychology of William James.* Pittsburgh, PA: Duquesne University Press.
Mandler, J. M., & Mandler, G. (Eds.) (1964). *Thinking: From association to Gestalt.* New York: Wiley.
Mayer, R. (1983). *Thinking, problem solving, cognition*. New York: W. H. Freeman and Co.
Mohanty, J. N. (1972). *The concept of intentionality*. St. Louis, MO.: Warren H. Green Co.
Runes, D. D. (1958) *Dictionary of philosophy.* Ames, IA.: Littlefield, Adams and Co.
Sartre, J.-P. (1956). *Being and nothingness.* (H. Barnes, Trans.). New York: Philosophical Library. (Original work published 1943)
Spiegelberg, H. (1982). *The phenomenological movement* (3rd ed.). The Hague: Nijhoff.
Titchener, E. B. (1898). The postulates of a structural psychology. *Philosophical Review*, *7*, 449–465.
Titchener, E. B. (1907). *A primer of psychology*. New York: The Macmillan Co.
Titchener, E. B. (1921a). Brentano and Wundt: Empirical and experimental psychology. *American Journal of Psychology*, *32*, 108–120.
Titchener, E. B. (1921b). Functional psychology and the psychology of act I. *American Journal of Psychology*, *32*, 519–542.
Titchener, E. B. (1972). *Systematic psychology: Prolegomena*. Ithaca, New York: Cornell University Press. (Original work published 1929)
Watson, J. B. (1913). Psychology as the behaviorist views it. *Psychological Review*, *20*, 158–177.
Woodworth, R. S. (1934). *Psychology*. New York: H. Holt and Co. (Original work published 1921)
Wundt, W. (1902). *Outlines of psychology* (C. H. Judd, Trans.). New York: G. E. Stechert. (Original work published 1901)
Yerkes, R. M. (1911). *Introduction to psychology*. New York: H. Holt and Co.

CHAPTER 4

CAN AN EMPIRICAL PSYCHOLOGY BE DRAWN FROM HUSSERL'S PHENOMENOLOGY?

BARBRO GIORGI

INTRODUCTION

This chapter will seek to explore some of Husserl's ideas about consciousness that are helpful to contemporary research. A descriptive phenomenological psychological research method was founded by Amedeo Giorgi (Giorgi, 1985) which is based on Husserl's philosophical method and which will serve as the framework for the discussion in this chapter. There are numerous publications already in existence that deal with many aspects involved in the method (Giorgi, 1970, 1981, 1983, 1992, 1994, 2000), however, this chapter will attempt to cover the manner in which some of Husserl's concepts are implicitly involved not all of which have received sufficient articulation elsewhere.

Most major phenomenological philosophers believe that phenomenological philosophy could help psychology in various ways. For example, in 1925, Husserl himself gave a course on phenomenological psychology (Husserl, 1962/1977). Merleau-Ponty (1942/1963, 1961/1964) also wrote extensively about the relationship between phenomenology and psychology. However, these analyses were conceptual and philosophical and do not clarify exactly how one would practically apply the phenomenological approach in psychology. The first person to attempt a rigorous practical application of the phenomenological method as developed within the continental philosophical tradition was Amedeo Giorgi. He took the philosophical concepts presented by Husserl as well as Merleau-Ponty (Husserl, 1913/1983; Merleau-Ponty, 1945/1962), modified and adjusted them in such a way that they would be useful to scientific purposes rather than philosophical ones. Cloonan (1995) has provided an extensive history of this development of phenomenological psychological research prior to Giorgi as well as a description of Giorgi's contribution. Thus, I will only briefly go into the history of the development of the method here.

Giorgi's background was in traditional experimental psychology, and he wanted to bring the same degree of success to psychology that the natural science method brought to nature. However, his experiences with the natural science method in psychology made him to realize that the shift of focus from "thing and process" to "humans and relationships" was more pervasive and radical than most psychologists seemed to realize. Moreover, having been exposed to phenomenological philosophy, he realized that a different perspective on science had to be developed.

Giorgi was aware of the "American grassroots" phenomenology of Snygg (1941) and Combs and Snygg (1959) but it did not seem to him to have appropriate philosophical backing. The same was true of Rogers' (1964) approach to phenomenology, and in addition, the latter did not use phenomenology in a methodological way. Finally, Giorgi was aware that both MacLeod (1947) and van Kaam (1959) promoted the phenomenological viewpoint from a continental perspective, but neither applied the insights to formulate or found a radical method. MacLeod (1949) was against such an idea in principle because he considered phenomenological analysis to be propaedeutic to science, and after such an analysis, one would use traditional experimental methods. van Kaam (1959, 1966) restricted his research to purely theoretical work which was also not necessarily restricted to phenomenological criteria. Consequently, after spending the greater part of the 1960s searching for a full-fledged psychological phenomenological practitioner with an articulated method, or published examples of such, and finding none, Giorgi went to philosophical sources themselves. The primary sources he used to develop a scientific application of the phenomenological method were Husserl's (1913/1983) *Ideas I* (pp. 171–184) and the Preface of Merleau-Ponty's (1945/1962) *Phenomenology of Perception.* Based on these writings which present a philosophical method, using imaginative variation and trial and error, Giorgi came up with a restructured method with a different procedural order which resulted in a phenomenological scientific research method adapted for conducting research in psychology.

The ways in which some of the phenomenological concepts had to be modified from their philosophical application in order to be useful for the scientific application in psychological research will also only be touched upon here, where necessary, since this step has received treatment elsewhere (Giorgi, 1985, 1989b, 1997). Instead, I will begin with a brief presentation of the actual steps of the descriptive phenomenological psychological method as articulated by Giorgi (1985) and then give a more detailed and lengthy elaboration of the way in which Husserl's phenomenological concepts are actually being employed in the implementation of the method.

THE SPECIFIC PROCEDURES OF THE METHOD

The method is executed by following four different steps which will be presented in turn.

Step 1—Reading for a sense of the whole.
Step 2—Dividing into meaning units.
Step 3—Transforming the data.
Step 4—Synthesizing the transformed meaning units (Describing the structure).

The research always begins with a description of an experience from an everyday perspective that is to be understood psychologically. The description, more often than not, is obtained by means of an interview. The purpose of the interview is to have the participant describe in a faithful and detailed manner an experience of a situation that exemplifies the phenomenon that the investigator is interested in. In other words, one could be interested in shyness, anger, joy, frustration, anxiety, violence, depression, or whatever, and the participant's role as an ordinary person from the everyday world, is to describe a situation in which he or she experienced such an event. The assumption is that the describer is not especially psychologically sensitive or focusing in particular on the psychological significance of the event. The transcription of the interview then becomes the raw data of the research. Once the researcher has the description, the following steps constitute the analysis.

It may be helpful to the reader to appreciate that each step of the method is a finer and more particular analysis built on the previous step, until the fourth step, which is once again a holistic and summary articulation of the phenomenon.

STEP 1: READ FOR A SENSE OF THE WHOLE

The entire description has to be read because the phenomenological perspective is a holistic one. One cannot begin with an analysis of a description without first having understood the whole situation. That is the major point of this first step, one does not analyze what one is reading or attempt to organize it in any way. Rather, one tries to "get into the story," allowing one's own subjective response to the whole situation to emerge. One needs to know the overall sense of the description before continuing on to the next step.

STEP 2: DIVIDING THE DESCRIPTION INTO PARTS: ESTABLISHING MEANING UNITS

Because most descriptions are too long to be handled in their entirety, parts have to be established in order to be able to achieve a more thorough analysis. Since clarifying the psychological meaning of experience is the goal of the analysis, it makes sense to partition the data into parts that are based on a psychological criterion, and the resulting parts are therefore called "meaning units," because they are demarcated on the basis of perceived transitions in psychological meaning. In other words, this division of the data is accomplished from within the perspective of the phenomenological reduction and with a psychological attitude, mindful of the phenomenon being researched. It has to be appreciated that there are no "objective" meaning units in the description as such. Dividing the description into meaning units is simply a pragmatic research tool and the meaning units are not theoretically weighty, merely helpful in the analysis. The only guide to "good versus bad" meaning units is that if the meaning units are too short one risks fragmenting the description, if they are too long the risk is that important aspects are glossed over and missed.

Step 3: Transformation of Meaning Units into Psychologically Sensitive Expressions

The third step involves transforming the concrete expressions in each meaning unit into the psychological meaning of those expressions. Specifically this means taking what is described both explicitly and implicitly and asking, from a psychological perspective, "what does this mean?" The data give a lived account of an experience rather than an analysis of the experience and the task of the researcher is to articulate the psychological meaning exemplified by this particular account.

As mentioned, the analytic work follows a progressive refinement in terms of its psychological sense from the original description through to the final outcome of a general structure. In the first step one reads what the participant expressed and, without analyzing its meaning, one simply responds to the whole description. Then the next step is to divide the description into parts and to make meaning discriminations that are psychologically relevant to the phenomenon being researched. The third step, which is the core of the analysis, expresses very directly the psychological meaning embedded in the participant's everyday experience. The whole purpose of the method is to discover, articulate and make explicit the psychological meanings being lived by the participant that reveal the essential nature of the phenomenon being studied. The original description is full of "everyday expressions" and it is full of references to the participant's specific world. The meanings directly and indirectly expressed by the participants have to be made psychologically explicit vis-à-vis the phenomenon under investigation as opposed to how it pertains to the specific participant in his or her personal life. This then means that personal meanings are investigated not for their own sake but for the value they have in clarifying the context in which psychological phenomena manifest themselves.

In this transformation of the data from the participant's everyday language and expression into a psychological language that more directly expresses the psychological meaning, the use of psychological jargon as it exists in the literature should be avoided. Because no theoretical perspective is as broad as the psychological perspective as such, and the language used in different schools is embedded with various assumptions that may or may not be accurate, it is safer to stay away from theoretical jargon whenever possible. What is being sought after is the psychological perspective of the practitioner, and not a specific theoretical perspective such as psychoanalytic or cognitive ones. What is being advocated here is the adoption of a generic psychological perspective as opposed to, say, a sociological or anthropological perspective. Theoretically speaking, the articulation of a discipline-wide psychological perspective has not yet been formulated or accepted, however, many clinicians adopt a theoretically "eclectic" or neutral perspective in their concrete work. This is where the general practitioner dwells, except for those who make a point of positing a theoretical position, and the phenomenological method requires the same general attitude. However difficult it may be, ordinary language that captures the psychological meaning sensitively and with appropriate complexity and nuances is what is sought. Mere labeling that does not articulate the actual psychological meaning should also be avoided.

STEP 4: SYNTHESIZING THE TRANSFORMATIONS INTO A STRUCTURE

The previous step of the analysis ends with a series of transformed meaning units. That is, meaning units that were originally in the language of the participant are now expressed with heightened psychological sensitivity in terms of the phenomenon being studied. In the fourth step, one then uses imaginative variation on these transformed meaning units in order to see what is truly essential about them and then one carefully describes the most invariant connected meanings belonging to the experience, and that is the general structure. Usually, it is the relationship among meanings that constitutes a structure. Using one's imagination one varies the facts contained in the description either by removing them or by changing them in various ways to see if the phenomenon and the embedded meanings stay intact or whether the variation drastically changes the phenomenon. If the phenomenon and its meanings stay intact even with the altered facts, then those facts are not essential to the phenomenon. If by contrast, some critical meanings are changed due to the variation, then that aspect of the description that was changed is indeed essential. Meanings would have to be expressed in such a way that all variations are encompassed.

This is a very brief summary of the method and what will follow is an elaboration of the phenomenological concepts that are explicitly and implicitly used in the method. Readers interested in a more elaborate description of the method, in some examples of analyses or in further theoretical articulations may find these in the following sources (Giorgi, 1985; 1986; 1987a; 1987b; 1989a; 1989b; 1992; 1994; 1997; Giorgi & Giorgi, 2003a; 2003b).

HOW PHENOMENOLOGICAL CONCEPTS GUIDE THE RESEARCH

What follows is an elaboration on the way in which some key phenomenological concepts are guiding the entire process of research. One could follow the above steps non-phenomenologically. Below will be articulated certain non-methodological concepts that will ensure that the attitude remains phenomenological. Many of the concepts are implicated in various aspects of the research process and an attempt to outline how the concepts are used either directly or indirectly will be made.

THE LIFEWORLD

Psychological research is continuously criticized for a lack of relevance, or lack of ecological validity, particularly in terms of clinical and therapeutic research (Chestnut, Wilson, Goldfried, & Wolfe, 1996; Mahrer, 1997; Seligman, 1996; Wright, & Zenlich 1987). The relevance controversy raises the issue of whether the conditions demanded by the traditional paradigm allows for data collection and analysis that is close enough to the actual conditions of the practice of therapy so as to shed significant light on the therapeutic process. Can the "scientific method," as traditionally practiced, ever sufficiently approximate the actual practice of therapy to truly clarify the psychological meaning or dynamics of therapy? The limits that traditional researchers experience force them to raise the question of whether there may not be better strategies for research into situations where human relationships are primary. Goldfried and Wolfe (1996, p. 1010) state that:

> The medical model of outcome research, with its emphasis on disorders and their symptoms in current clinical trials, also has the particular limitation of neglecting the determinants/dynamics that clinicians know well to be essential to the change process.

They go on to say (1996, p. 1013):

> . . . our current research paradigm will neither provide us with all of the information that a clinician will need, nor will it substantially close the gap between research and practice. What we need is an alternative research paradigm for building and testing an effective approach to psychotherapy, one that both emerges from therapist-client interaction and individualizes the intervention for the particular case at hand. The implication of such research is that what needs to be specified and replicated is not brand-name therapies but identifiable processes of client change and the therapist behaviors that bring these about. The need to develop a research paradigm that also individualizes the intervention on the basis of an initial assessment and case formulation is essential for closing the clinical-research gap.

Seligman (1995, p. 966) wrote: "The efficacy study is the wrong method for empirically validating psychotherapy as it is actually done, because it omits too many crucial elements of what is done in the field."

Husserl's concept of the lifeworld is very helpful in regards to these concerns. What is meant by lifeworld is to turn to examples of the way things are actually lived and experienced in the context and situation in which they occur. In other words, when an experience of a phenomenon is investigated it is done via real life examples in which the phenomenon or experience is embedded rather than creating a battery of questions pertaining to specific isolated aspects of the phenomenon which are removed from the full context in which these aspects where experienced. By going to the lifeworld to collect the raw data, phenomenology meets the objections voiced by traditional researchers. The problem of the research situation being too far removed from praxis is eradicated when the raw data consists of concrete descriptions of lived events by the individuals living through the experience. Phenomenology matches the desire for a new paradigm that can offer the kind of research findings that would be more useful to the practitioner. An important part of the problem stems from the traditional natural science paradigm seeking to isolate relevant variables and examine them in isolation so that claims may be made concerning those variables. In order to do so, psychological constructs and operational definitions are created such that variables can be isolated and examined. But, in doing so, fidelity to the phenomenon as it is lived and experienced is lost. Although a logically coherent and often persuasive pattern emerges from studying well organized and controlled constructs, the patterns that emerge do not sufficiently fit human beings as they live and experience their lives. Variables can be mentally or intellectually separated and seemingly isolated and then relationships among them sought. In lived experience, however, what we isolate is already related and must be understood as already inevitably related. "Going back to the things themselves" means capturing things as they are lived and avoids this problem of lack of relevance and ecological validity. Collecting data from the lifeworld of the participants rather than attempting to "objectively measure" or quantify the data is one important aspect of doing phenomenological research that insures the desired faithfulness to the phenomenon. Meanings arise from patterns of behavior or conscious acts and not through isolated variables which is also why phenomenological psychologists describe their results in terms of general structures. It is the network of meanings that count, not their isolated sense. Collecting the data from the lifeworld means that the phenomenological findings are not based on constructs, nor does such a procedure lead to the creation of themes, categories, or indices of behavior. Findings are not based on isolated variables, but are descriptions based on essential meanings and structures of situated and contextual constituents. The phenomenon can be comprehended exactly as the participant experienced it, with all the nuances and shadings required. Moreover, since there are no structural or time requirements for phenomenological interviews, the participant and the researcher can go into whatever depth they are capable of reaching. Neither complexity, flexibility nor variability is a problem because it can be described exactly as it took place. For phenomenology, a phenomenon is that which shows itself to the consciousness of the perceiver, precisely as it is given. Perhaps a simple way of summarizing this is to say that efficiency is not a strong value for phenomenology. Rather, fidelity to the phenomenon being researched is what counts.

The lifeworld is also indirectly implicated in a more subtle sense in the interview situation in that the open-ended interview is faithful to the phenomenon of "gathering a specific kind of data." The open-ended interview is close to "the type of relationship" in which sharing one's own experiences with someone else naturally occurs in the lifeworld. Within this kind of relationship, the interviewer must be an approachable human being since more and better data is likely to be provided in a situation where there is a sense of connection and rapport with another person. The more psychologically sensitive the topic of the research is, and the more difficult it is for the participant to talk about, the more important the relationship between the participant and the interviewer becomes. An intimate conversation about sensitive material with an approachable and safe other is more likely to produce a "faithful description" of the participant's experience than would a contrived research situation such as a pen and paper measure or a pre-determined interview with specific questions to answer. In the lifeworld, intimate personal material is not shared in a "testing situation" but requires an authentic relationship. In the lifeworld, personal material is shared in a context of intersubjectivity. In an interview where the rapport is good and the interviewer is listening to the participant at the intersubjective level, the participant will feel listened to, understood but not forced into any direction that is not truly his or her own and will be more able to open up and describe the experience at an authentic level. This leads us to the concept of intersubjectivity.

INTERSUBJECTIVITY

It seems as though that, under the banner of science, researchers believe that some form of objectification or quantification is necessary. Traditional researchers believe that in order to achieve objectivity, one must objectify. Objectification means that one produces some form of external index of a process, thus giving a public manifestation of a lived event that is equally accessible to all, but which is more often than not lacking in faithfulness to the phenomenon as experienced. Objectivity in the proper sense means capturing an event accurately, "as it really is." Often this requires intersubjective cooperation and shared meanings that depend on the other being able to assume the appropriate subjective attitude to achieve the objective or accurate outcome. In other words, the objectification process achieves public commonness but at the price of fidelity. To achieve this public commonness without losing fidelity, intersubjectivity rather than objectification is necessary.

Husserl's (1962/1977) distinction between intrasubjective and intersubjective is of great help to us here. Each one of us has our own personal experience based on our personal history, our individual characteristics, style, preferences, etc. This is our intrasubjective experience. However, in order to understand and relate to each other we must be able to extend beyond our own intrasubjective or idiosyncratic experience to a level of intersubjectivity. We do this often and spontaneously. A simple example is "having fun." I can relate to and understand the experience of having fun that you describe to me because I also have that experience while I cannot understand how you can have fun through this particular activity which I personally find boring. If I used my intrasubjectivity I would not be able to understand your experience of having fun, but if I relate at the intersubjective level I remove my own personal details (the particular activity) and relate to your experience in a more "generalized human" manner and I can now understand your experience of having fun. This distinction between intra and inter-subjectivity clarifies the issue between the desire of science to achieve "objectivity" rather than "mere subjectivity." Indeed, intrasubjectivity is not desirable for a scientific understanding, this is indeed "merely subjective," but intersubjectivity it what is needed rather than objectivity.

Although it is clear that what is needed in the context of research is intersubjectivity, the issue of intersubjectivity also contains a certain tension, it requires a balance between the intrasubjective and the intersubjective which is critical and sometimes difficult to achieve but is nonetheless possible. The researcher must have a certain degree of intersubjective understanding of the cultural, historical, social context in which the participant is embedded both in regards to the interview as well as the data analysis. This is also true in terms of the psychological implications of the phenomenon itself. The researcher must have some experience of his or her own in order to relate to the participant's experience. Without sufficient intersubjectively shared meanings, the researcher cannot gain insight or derive the meanings experienced and expressed by the participant. So, the researcher must have personal experiences similar enough to the participant's in order to achieve the required intersubjective understanding. For example, conducting research within a culture where one does not understand the customs, symbols, and meaningful relationships means that one will understand the description according to a different set of rules than the ones used by the participant. The need for this kind of insightfulness is clearly understood in anthropology. Where psychology is concerned, however, the researcher must in addition to this have some kind of psychological experience to fall back on in order to fully see the psychologically embedded meanings. If one never personally experienced any kind of fear, one would not be able to see the embedded meanings in a description of a fearful event. This, of course, is where the critical balance between the intersubjective and intrasubjective comes in. One might say that the more the personal experience of the researcher and the closer the researcher is to the phenomenon and the context in which it occurred, the better the researcher is able to see the implicit and embedded meanings. This is also undoubtedly the biggest threat to the reduction. The closer the researcher is, the more difficult and the more important it is to employ the reduction and not be influenced by these same personal experiences and understandings. It is not an easy task to listen to the participant's data from a perspective of shared meanings while at the same time not allow one's own perspective on those familiar aspects to influence one's listening to the participant's unique experience of them. This balance is no doubt impossible to obtain with any kind of perfection, but at least, the distinctions among intrasubjective versus intersubjective and objective versus objectification mentioned above, resolves the issue of what constitutes scientific knowledge mentioned above. One cannot access the embedded meanings lived and expressed by the research participants from a detached, non-subjective objectifying stance. It takes another human subjectivity to relate to the subjective experience of another human being. This distinction made by Husserl (1962/1977) helps us see that neither objectification nor insufficient intersubjectivity will allow access to the data at the truly psychological level. That does not, however, mean that we are relegated to the "merely subjective" which would not lead to objective scientific knowledge. Intersubjectivity, rather than reified objectification, leads to a faithful understanding of the phenomenon.

A related problem is the issue of direct experience which is a phenomenological requirement. The outcome of the analysis is, as we have seen, based on the psychological meaning discovered and made explicit by the researcher, while these may not be explicitly stated as such by the individuals having the experience. Thus, the final essential structure is a description of the psychological meanings that, from the original participant's description, through the researcher's analysis, become present to the consciousness of the researcher, thus fulfilling the phenomenological requirement of using only direct experience. This also means that the critical check of the original researcher's procedure can be performed by any competent colleague who has sufficient access to the intersubjective arena in which the events described occurred and the required phenomenological theoretical knowledge. This then, based on Husserl's idea of our human capacity to respond to that which presents itself to us in direct experience, offers a public manifestation of a lived event that is equally accessible to all without losing the fidelity to the phenomenon through objectification.

INTENTIONALITY

Husserl (1962/1977) suggests that one of the basic ways in which our consciousness functions is that consciousness is intentional. What this intentionality means is that consciousness reaches out towards something that transcends it and this reaching out beyond itself always has an object towards which it is directed. For example when we feel something such as love or anger, it is always directed towards something or someone, when we think it is always about something or someone, when we see or hear there must always be an object for this seeing or hearing act of consciousness. An important contribution to the task of research lies in Husserl's understanding of consciousness as not limited to the dimension of awareness. It may be safe to say that most of consciousness actually happens outside of our immediate awareness but whatever the ratio of conscious versus unconscious material may be, it is critical to understand that Husserl's notion of consciousness includes the entire spectrum from fully conscious to fully unconscious material (Husserl, 1962/1977). This notion is helpful to us because it permits us to look at events in the lifeworld and discover that which is implicitly embedded in the situation without being limited to the part of experience that lies within awareness. Given that the structure of intentionality always has an object whether this object is in our awareness in a clear and distinct way or out of awareness and only indirectly experienced, the researcher can intuit meanings that are embedded in the experience without them being explicitly expressed by the participant and hence make them explicit. The vehicle for this process is our human ability to intuit "the things themselves," to intuit that which presents itself to consciousness directly.

INTUITING

Husserl's notion of intuiting or intuition is also helpful to us in the sense that it broadens our usual response to the concept. Intuition is rarely a word that appears in the context of research because we tend to interpret intuition as vague responses we may have in situations where we have inadequate or no factual or informational basis. This is not Husserl's meaning of intuition. Husserl includes in this concept all levels of "knowing" from the simple factual observation that there stands another person in front of me, to the more subtle aspects such as "I have a hunch" that this person is anxious about something. This incorporation and acknowledgment of the whole spectrum of human intuition, whatever is present precisely as present, legitimizes the more subtle levels of intuition (usually excluded from scientific enquiry), empowers the researcher tremendously in accessing that which is not easily and directly observable. The difficult challenge here is to discriminate between what is "just a hunch" and what is legitimate subtle intuition based on subtleties in the data. When the researcher intuits an embedded meaning, the task is to articulate the subtle information within the data that produced this intuition. Husserl's opening up and expanding for us the understanding of our own intuitive capacities allows or even forces us to look closely at what presents itself to us. A research situation then demands of the researcher a careful tracing back to what, in the data, carried with it the information that resulted in what presented itself to the consciousness of the researcher. Important for psychology is the fact that behavior is also intentional, i.e., directed to situations that transcend the behavior itself, and using human intersubjectivity we have the capacity to intuit these intentional objects. The mode of consciousness that is directed toward objects also influences how an object appears—thus whether an object is grasped as desired, as feared, with certainty, questioningly, etc., will have an impact on the sense of the object. Therefore, since it is the sense of the object or state of affairs that phenomenology seeks, the raw data should contain as much of the mode and manner of consciousness as well as the "whatness" of the object so that the researcher can intuit and draw out the full meaning of the experience.

PRE-REFLECTIVE

A point that follows from the previous discussion also has to do with the kind of description that is used in the descriptive phenomenological research method. Participants are always asked to describe a concrete event as an example of the experience being investigated. Participants could, of course, describe their experience in terms of what they think about it along with how they feel about it. This, however, is not what is sought. How the experience is thought about is, of course, part of the experience and should be included, but it is only a small portion, and not the important part of the description. A good description is one that describes what took place, what happened, and that includes the intellectual and emotional context in which the event took place in as rich a detail as possible. The reason for this is based on the understanding of the critical influence of the pre-reflective as opposed to the reflected upon and understood aspects of human experience (Merleau-Ponty, 1961/1964). Most of the experience of our life does not take place in the realm of what we know we are doing, feeling, or thinking. We respond to our world before we reflect upon it and the reflections we do make are ultimately only a small portion of the actual experiences of, or responses to the world. This has far reaching implications for research in psychology. As a data source, the concrete description is the richest possible source of psychological data because it does not require the participant or the researcher to "know" or understand what is taking place in regards to the phenomenon. If we asked the participants for their understanding of their own experience this would severely limit what the participants could tell us about their experience. However, by asking participants to describe "what happened." they can inform us about their experience from their full human capacity of pre-reflective responding to the world rather than the more limited area of reflection. The reflections the participant has made would be included in a description about what happened because that is usually perceived by the participant as necessary contextual information in order for the interviewer to understand the meaning of the event.

A second way in which the limits of knowing enter into the research situation is from the perspective of the researcher. Traditional research requires questions to be asked by the researcher. This may be a serious problem because the questions must be the right ones, thus requiring the researcher to know in advance what constitutes relevant and necessary information in order to shed accurate light on the phenomenon under investigation. Traditional research first of all requires a choice in regards to what should be included in terms of topic, time, context and others. Let's take a concrete example to demonstrate how difficult these choices can be. Let us say that we want to research "violence among the youth" given the recent incidences of students coming to school with guns and shooting other students and teachers. In terms of the topic, what should we ask? Should we ask about feeling: angry, frustrated, hopeless, helpless, rage, entitled, abandoned, victimized, ridiculed, lonely, shame, guilt, fear, anxiety, tension? Should we ask about personality characteristics or actions: manipulative, in denial, depression, avoidance, challenging behaviors, aggressive behaviors, irresponsible, stealing, lying, trying to impress, seeking approval, monitoring behaviors? What about time frame, do we include questions that cover the period from childhood to this moment and projections into the future or do we set some form of time limit? Then there is context. Do we investigate the school environment, and if so in what sense? Classroom size, amenities, extra curricular activities, student/teacher ratios, number of students in school, ethnic mix of students and teachers—all these are possible. What about the media, TV, movies, music, papers, magazines, and if so how detailed and in what sense? Then there is culture and ethnicity, socio-political position, religious belonging, belief system both individual and socially/culturally embedded, and so forth. What do we need to include and how detailed must the investigation be? Last, but not the least, what others must be included? Do we have to include all significant others, parents, siblings, friends, teachers, community personnel (priests, coaches, therapists, vendors, hair dressers, restaurant personnel, etc.) role models and heroes (movie star, sport stars, etc.) or can we know that all these people are not important? If we want to trace what the various influences are and signposts for the problematic behavior of one day taking a gun to school and shooting friends and teachers, what do we need to include in our investigation in order to truly understand what happened from a psychological perspective? These are the kinds of choices necessary for the traditional researcher and the task of making the right choices is daunting. It presupposes a great deal of knowing in advance what is relevant and what is not, the way in which it is relevant and the level of detail that is necessary and meaningful. One of the pitfalls of research is to get lost in our understanding of matters according to the way we have learned to conceptualize them. Allowing the participants to express themselves about their own experience without the influence of the researcher's question sidesteps these problems. If a research participant is asked to describe an actual event that contains the phenomenon under investigation, then the participant will spontaneously contextualize the event and give the most relevant and important aspects of the experience. When we tell someone about something that happened in our life, we spontaneously give the required background to the person so that they will be able to appreciate what actually happened. This includes the various topics that are most relevant, the time frame required for understanding and the context in which it took place. Going back to our example of a psychological investigation of school shootings, if we could interview someone who actually did this about the experience in a richly detailed and honest interview, he or she would tell us the relevant contextual background for understanding the meanings of the event rescuing the researcher from having to predict in advance what should be included. The participant would choose the time frame for the beginning of the story, who needed to be included, etc., in order to make sense of the event itself. This not only avoids the perilous situation of having to make choices about what to include but it also opens the door for surprises and discovering things that could not have been foreseen.

We already established that the influence of the interviewer is important both in terms of the ability to enter the proper intersubjective realm and in terms of the interpersonal relationship required for the participant to be willing and able to share their experience with the interviewer. This, from a research perspective, may also be problematic. We do not want research findings that are a reflection of, or colored by, the researcher. Here again, the description of a concrete event is helpful. Because the description is of an event that the participant lived through, there is less influence by the researcher since the event itself did not involve the researcher. Although there is some influence by the interviewer as to what and how much of what happened will be revealed, at least "what happened" is independent of the interviewer and the relationship during the interview. In a comfortable interview situation, the participant should be able to "tell it like it happened" and not reformulate the event for the sake of the interviewer. To the extent that there is a desire on the part of the participant to give a certain self-presentation, this is also much more difficult to achieve in a description of a lived event and what took place than it is when giving an intellectual description of what we think of something. The control over and choice of what to reveal is much greater in an intellectual summary in regards to a phenomenon than in a description of an actual lived experience of the phenomenon. To summarize, the phenomenological method is descriptive—not interpretive, theoretical or intellectual because its point of departure consists of concrete descriptions of experienced events from the perspective of everyday life by participants, and then, the end result is a "second order" description of the psychological essence or structure of the phenomenon as experienced by the consciousness of the researcher.

The Epochē

When it is said that within the reduction everything that presents itself is to be accounted for precisely as it presents itself, it is a strategy devised to counteract the potentially biasing effects of past experience. We already touched on this in terms of data analysis in the discussion of the tension between the intersubjective and the intrasubjective saying that in order to reach an appropriate intersubjectivity the intrasubjective must be involved but then bracketed so as not to influence the analysis of the data. Here, I would also like to point to the ways in which the epochē enhances the interview situation. At the more obvious level, bracketing is employed in the refraining from asking the participant a number of questions but instead letting the participant choose what to talk about from their own experience without interference from the researcher. Although the researcher may have ideas or hypotheses that he or she would want to verify with the participant, may wonder whether this or that took place, the researcher brackets these ideas or questions in order not to bring his or her assumptions to the experience of the other. In addition to this restraint, however, there is a more subtle level in which the attitude assumed in bracketing helps the interview situation. The researcher needs to not simply refrain from asking questions because the method requires this, but needs to have an authentic openness to the other which is an attitude embedded in the bracketing. This openness constitutes the ability to listen to the person and truly hear what they are expressing. It constitutes an openness towards the other that creates a space where the participants can feel free to express themselves in their own way and feel that this is what is sought and that no judgments or conditions are placed upon what their experience ought to be. This attitude is very similar to that of the therapist who must be able to listen to the client openly and acceptingly without judging or imposing his or her own views on the client.

It should be pointed out here that the phenomenological reduction used in psychological research at the scientific level is different from the transcendental reduction of the philosophical level. In research we do not wish to accomplish a complete transcendental reduction but one that to a certain extent stays situational and contextual since psychological events or phenomena inevitably happen within a context and must be understood within that perspective. This scientific phenomenological reduction Husserl called the psychological phenomenological reduction (1962/1977). What this also means is that the objects or states of affairs experienced are reduced, but not the acts of consciousness with which the objects or states of affairs are correlated. It means that the acts are taken exactly as they present themselves—as belonging to an existing psychological subjectivity. However, the objects correlated with the acts are taken up without making an existential claim. That is, what is experienced is understood to be an experiential given to the person experiencing the object, the person is genuinely experiencing some given phenomenon, but the claim that what is present to the person's consciousness actually exists the way it is given is not claimed by the researcher. This lack of pressure towards an existential truth claim can also have a subtle influence in the interview situation. When a participant describes his or her experience to a researcher who is listening to the described experience from the perspective of "this is how this was experienced by you and is therefore true for you" it creates an openness that invites the participant to continue. A more realistic attitude toward what the participant is saying, i.e., a judgmental attitude that questions the participant's account (e.g., "if this really is true" or even worse, convinced that "this cannot be true"), can subtly influence the degree to which the participant will be able and willing to describe his or her authentic experience.

This understanding of truth is furthermore extremely helpful to psychology since much of our subject matter is exactly that which does not objectively exist. One may even say that often, the bigger the gap between the perceived reality and the objective truth, the more psychology is involved. To take a concrete example, if I am walking alone down a dark alley late at night and I hear footsteps that are getting closer and closer my sense of being followed along with an anxious response is more objective than the same sense of being followed with a concomitant anxiety when I am sitting in a classroom listening to a lecture. The latter situation is labeled "paranoia" and understood as a psychological/pathological condition exactly because it does not match the objective reality. Not focusing on objective reality as such but examining the lived experience as it is lived and taking that as a true presentation for the person who's experience it is, allows us to expand our field of inquiry. When presence is the criterion of what is given rather than some objective true criterion, we can access a wider spectrum of human experience because we are not limited to looking for something objectively true nor trapped into trying to find "where" something is located such as in this or that part of the brain. This particular point, of course, also involves Husserl's distinction between real and unreal objects.

In addition to this aspect of the reduction there is yet another way in which the reduction is involved in the research process. To utilize the epochē means to bracket past knowledge about the experienced object, in order to experience this instance of its occurrence freshly. When we encounter familiar objects we tend to see them through familiar eyes and thus often miss seeing novel features of familiar situations. This is what is meant when phenomenologists say that they want to experience things "freshly" or "with disciplined naiveté." Even if objects turn out to be precisely as we first thought, it is more rigorous to give nuances and "taken-for-granted" aspects a chance to show. In practice, it can be difficult for a researcher to look at some phenomenon that he or she has studied for years and has become an expert on, to still see with fresh, naïve new eyes at what presents itself. An interesting example of this comes to mind where a biological scientist recruited a number of young children to study the patterns of butterfly wings. The biologist had studied butterflies for years and was so used to looking at their various patterns that she lacked the freshness of the gaze of young children simply looking with curiosity at the butterflies. What Husserl suggests is that we must return to that which presents itself directly to our consciousness and look at it as freshly as possible. One can sometimes observe in children an uncanny ability to read situations that as they grow into adulthood weakens. This sort of directness in the reading of the situation is perhaps a good example of the kind of intuition that serves the psychological researcher well in terms of getting back to the things themselves.

Signifying, Fulfilling and Identifying Acts

In the transformation and analysis of the meaning units in step 3, one often goes back several times and revises the transformations until one is satisfied that all the psychological meanings contained in the meaning units have been articulated as completely and accurately as possible. The question may then be asked "How does one know when the analysis is complete?" This issue is greatly clarified by the implementation of an aspect of Husserl's theory of meaning, the idea of the signifying, the fulfilling and the identifying acts. Husserl suggested (1900/1970) that we posit an idea or a desire or a question of some kind (an intentional object) which sets in motion a search to locate or satisfy this signified object. In this search, we will turn our attention to objects that we see as possibly fulfilling the signified object and when we have located an object that fulfills the signifying object precisely we have reached an identifying act and the search will then stop. Let us take some concrete examples of this. Let's say that my signifying act is "I want a good book to read." I will then proceed to a book store or a library, books being fulfilling objects, but I will not stop looking until I find a book that looks like a good one to read at which point I found the identifying object and stop searching for more books. This is a human activity that goes on sometimes with great awareness accompanied by goal oriented behaviors, such as in the above example, other times we may be engaging in this search without awareness only to be struck by insights such as "This is what I was looking for all along, I just didn't know it." Another interesting example is Glen Miller who for a number of years was aware of looking for a certain sound. He experimented with various musical arrangements and his colleagues told him it sounded great but he kept insisting that it isn't the right sound yet, not the sound he was looking for. Eventually he found that sound, he knew he had found it and stayed with it from that point and we all now recognize that sound as the unmistakable "Glen Miller sound." Perhaps the trickiest area for most of us where this searching is involved is the signifying act of "I'm lonely, I'd like to find a partner." We may posit an idea of what this person should be like and proceed trying out one relationship after another, the fulfilling acts, until we hopefully find the one person who is right, which leads to the identifying act. Often this process is filled, time after time, with the hope that this person is the right one only to be disappointed and then one begins the search anew.

In the context of research, in the analysis of the description of a concrete event, the signifying act is finding an adequate articulation of the multiple meanings embedded in the examples of lived experience provided by the participant. The question becomes "what is the participant communicating psychologically by describing these events in this particular way?" As a human subjectivity, the researcher will have a sense of what this communication is, even if this is at first somewhat vague or ambiguous. In the work of the analysis then, the researcher starts to look for expressions and articulations that will fulfill his or her signifying act and when these second order descriptions fit and match the signified intentional object of the researcher, he or she will then have a sense of having reached a moment of identification and the search for additional or different ways of expressing the meanings in the description stops. There is a sense of having completed the task of transforming the implicit psychological meanings from the signifying acts as communicated by the participant in the description, a sense of capturing in the identifying act a complete and faithful articulation of what was implicitly embedded in the original description, thus making the psychological sense explicit. This may, to some, sound more like a matter of interpretation than a faithful description of the data. To be sure, the risk of making interpretations that belong to the researcher rather than to give a description of the participants lived meanings, is always present, but the use of imaginative variations is a very useful tool to combat this difficulty.

Imaginative Variation

Using imaginative variation as a tool means to use one's imagination to vary different aspects of the data at hand. If I, for example, want to make sure that the psychological meaning I assign a certain concrete event is not my interpretation but an accurate description, I will use imaginative variation. Let's say that I describe the participant's response to a situation as an "aggressive response." Using imaginative variation I can verify that this is not simply my interpretation by varying the factual details of the participant's description. Let's say that the description said something like "I decided to cut him off and make sure that he couldn't pursue his own interests but would be forced to do it my way . . . and I knew that this would really aggravate him." Varying the detail "and I knew this would aggravate him" to "but I knew he would appreciate this in the end since it would all work so much better in the long run for all of us if it was done my way" would change the psychological meaning from an "aggressive response." I could do similar variation to "cut him off" "be forced to," etc. This allows me to see that given the actual facts and expressions in the participant's description it is clear that it is an aggressive response rather than say a cooperative, friendly, constructive response. If this was not an aggressive response the factual details of the description would have to be different from what they are and therefore it is not simply my interpretation but indeed a description of the participant's response.

This method of applying imaginative variation is used both when transforming the meaning units in step 3 as well as in synthesizing the transformations into an essential structure in step 4. As mentioned above, there is always a risk that meanings which appear to the researcher may be the researcher's interpretation rather than the meanings described by the participant. The above example is one way to try to eliminate such interpretations where the researcher submits his or her articulations to imaginative variation to see if there indeed are other possibilities of articulating the meanings in the description. Another way of using imaginative variation to eliminate interpretation has to do with the amount of information available in the participant's description. Let's say that the participant describes someone laughing in response to a comment the participant had just made. If I did not have any further information, I could not describe any psychological meaning here since there would be several possibilities of meaning. I could say that the participant appreciated the support of the other's laughter, but by using imaginative variation, I would quickly see that I could imagine other possibilities as well. I can imagine that the comment was not meant to be funny and that the other had a generally critical attitude towards the participant, then the laughter could be experienced as a dismissal, a ridiculing of the participant. There is a distinct difference between laughing *at* or laughing *with* someone, and without further information I could not decide which this might be. If, on the other hand, the participant had described a situation in which the atmosphere was tense, the participant felt nervous and was scrambling for something to say to ease the tension and was also desperately looking for someone in the room to connect with and describes a sense of relief in a moment of eye contact with this other person who then laughed at the joke the participant delivered, then I could describe the experience as appreciating the support from the other's laughter. In this case, when I try to imagine that the participant felt ridiculed by the other's laughter it does not hold as a viable possibility. In this manner, imaginative variation is used throughout the analysis to make sure that the articulation of the psychological meaning is not just one possibility or interpretation among many, but is, by necessity of the circumstances described, a description of the participant's experience.

CONCLUSION

The challenges to conducting useful, meaningful and relevant research in psychology are many. It is my belief that we have paid a high price by accepting traditional natural science research methods as the one and only superior way of conducting research. However, thanks to the research alternatives offered to us by qualitative methods, and in particular phenomenology, there is hope for our discipline, over time, to resolve the crisis and bridge the existing gap between research and praxis. So much of what has traditionally been missed in terms of nuances and complexities can be discovered, addressed and articulated if we allow ourselves to be guided by the principles of phenomenology.

REFERENCES

Chestnut, W. J., Wilson, S., Wright, R. H., & Zemlich, M. J. (1987). Problems, protests, and proposals. *Professional Psychology: Research and Practice, 18*, 107.

Cloonan, T. (1995). The early history of phenomenological psychological research in America. *Journal of Phenomenological Psychology, 26*, 46–126.

Combs, A. W. & Snygg, D. (1959). *Individual behavior: A perceptual approach to behavior*. New York: Harper & Row.

Giorgi, A. (1970). *Psychology as a human science.* New York: Harper and Row.

Giorgi, A. (1981). Ambiguities surrounding the meaning of phenomenological psychology. *Philosophical Topics, 12*, (Suppl. to Phenomenology and the Human Sciences). 89–100.

Giorgi, A. (1983). Concerning the possibility of phenomenological psychological research. *Journal of phenomenological Psychology, 14*, 129–169.

Giorgi, A. (Ed.), (1985). *Phenomenology and psychological research*. Pittsburgh: Duquesne University Press.

Giorgi, A. (1989a). Some theoretical and practical issues regarding the psychological phenomenological method. *Saybrook Review, 7*(2), 71–85.

Giorgi, A. (1989b). One type of analysis of descriptive data: Procedures involved in following a scientific phenomenological method. *Methods, 1*(3), 39–61.

Giorgi, A. (1992). Description versus interpretation: Competing alternative strategies for qualitative research. *Journal of Phenomenological Psychology, 23*, 119–135.

Giorgi, A. (1994). A phenomenological perspective on certain qualitative research methods. *Journal of Phenomenological Psychology, 25*, 190–220.

Giorgi, A. (1997). The theory, practice and evaluation of the phenomenological method as a qualitative research procedure. *Journal of Phenomenological Psychology, 28*, 235–260.

Giorgi, A. (2000). The similarities and differences between descriptive and interpretive methods in scientific phenomenological psychology. In B. Gupta (Ed.), *The empirical and the transcendental: A fusion of horizons* (pp. 61–75). New York: Rowman and Littlefield.

Giorgi, A. & Giorgi, B. (2003a). The descriptive phenomenological psychological method. In P. Camic, J. E. Rhodes, & L. Yardley (Eds.) *Qualitative research in psychology: Expanding perspectives in methodology and design*. APA Publications.

Giorgi, A. & Giorgi, B. (2003b). Phenomenology. In J. Smith (Ed.) *Qualitative Psychology. A practical guide to research methods*. London, UK: Sage Publications.

Goldfried, M. R. & Wolfe, B. E. (1996). Psychotherapy practice and research: Repairing a strained alliance. *American Psychologist, 51*, 1007.

Husserl, E. (1970). *Logical investigations I and II* (J. N. Findlay, Trans.). New York: Humanities Press. (Original work published 1900)

Husserl, E. (1977). *Phenomenological psychology* (J. Scanlon, Trans.). The Hague: Nijhoff. (Original work published 1962)

Husserl, E. (1983). *Ideas pertaining to a pure phenomenology and to a phenomenological philosophy I* (F. Kersten, Trans.). The Hague: Nijhoff. (Original work published 1913)

Macleod, R. B. (1947). The phenomenological approach to social psychology. *Psychological Review, 54*, 193–210.

Mahrer, A. R. (1997). What are the "breakthrough problems" in the field of psychotherapy? *Psychotherapy, 34*, 81.

Merleau-Ponty, M. (1962). *The phenomenology of perception* (C. Smith, Trans.). New York: Humanities Press. (Original work published 1945)

Merleau-Ponty, M. (1963). *The structure of behavior* (A. Fisher, Trans.). Boston: Beacon Press. (Original work published 1942)

Merleau-Ponty, M. (1964). Phenomenology and the sciences of man. In J. Edie (Ed.), *The primacy of perception* (pp. 43–95) (J. Wild, Trans.). Evanston, IL: Northwestern University Press. (Original work published 1961)

Rogers, C. (1964). Toward a science of the person. In T. W. Wann (Ed.) *Behaviorism and phenomenology*. Chicago: Chicago University press, pp. 109–133.

Seligman, M. E. P. (1996). Science as an ally of practice. *American Psychologist, 51*, 1072–1079.

Snygg, D. (1941). The need for a phenomenological system of psychology. *Psychological Review, 48*, 404–424.

van Kaam, A. (1959). Phenomenological analysis: Exemplified by a study of the experience of "really feeling understood". *Journal of Individual Psychology, 15*, 66–72.

van Kaam, A. (1966). *Existential foundations of psychology*. Pittsburgh: Duquesne University Press.

CHAPTER 5

DID HUSSERL CHANGE HIS MIND?

An Epistemological Analysis That Connects Husserl's Philosophy with His Followers

KARIN DAHLBERG

INTRODUCTION

Husserl's notion of transcendence has been and continues to be the source of much philosophical discussion and dispute among human science researchers. As long as there has been a phenomenological tradition there has been debate as to whether or not to follow Husserl's ideas about transcendental phenomenology. Typically, such discussions easily become debates between extreme and opposing positions. One viewpoint sees transcendence as a notion of standing aside from one's subjective experience, in order to observe the world or a particular phenomenon from a pure epistemological perspective, one of total objectivity. One might imagine a free-floating platform upon which the phenomenologist sits and that provides an unhindered view of the phenomenon in question. People defending this position put forward the notion that the reduction is a necessary part of a methodological approach that aims to be scientific. The other viewpoint holds that such a position of pure transcendence is impossible and that the transcendental idea was built on a false understanding. These critics deny the possibility of a pure consciousness. They emphasize that transcendence should be considered a hypothetical philosophical notion, which therefore ought to be discarded altogether. In the wake of this dispute is the debate about whether Husserl's well-known followers Heidegger, Merleau-Ponty, and Gadamer reject his idea of transcendentality or not. It is obvious that Husserl's lifeworld theory became a substantial gift to subsequent philosophy, but whether or not Heidegger, Merleau-Ponty, and Gadamer also took up his transcendental phenomenology is more veiled.

Phenomenology is an important foundation for human science research and practiced in many research contexts beyond psychology, such as pedagogy, sociology, and health sciences (hence my use here of the more general term "human science"). Consequently, scholars who develop and practice phenomenological philosophy in order to conduct human science research have to address the idea of transcendentality within phenomenology. There are two main questions:

1. Is there a phenomenological coalescence or are there fractious, conflicting phenomenologies?
2. Is there a total break between Husserl and his followers due to the transcendental idea?

The purpose of this paper is to discuss the meaning of Husserl's phenomenology. I want to put forward a notion of the Husserlian transcendental theory, different from and going beyond the two extreme positions mentioned above regarding this notion. I want to show that Husserl was an equilibrist, aiming at balancing between the richness and ambiguity of the lifeworld and the scientific claims of rigor. I also propose that his major followers are not entirely ice-cold to the transcendental discussion in his philosophy, though this is often maintained.

PHENOMENOLOGY AS "GOING BACK TO THINGS THEMSELVES"

Husserl's main motives for developing a phenomenological philosophy were probably several. One of his motives was his observation that the tremendous progress of the natural sciences, including a burgeoning technology, lead to a governing ideal in science. The social sciences were especially, and detrimentally, influenced by naturalism and positivism. Husserl warned that cultivating the scientific ideal of naturalism and positivism would sever science from the everyday world. Even if this was written quite late in Husserl's life, it is obvious that it was a strong conviction of his, that

> merely fact-minded sciences make merely fact-minded people. The change in public evaluation of science was unavoidable, especially after the war, and we know that gradually it has become regarded with hostility among the younger generation. In our vital need, so we are told, this science has nothing to say to us. It excludes in principle precisely the questions which man, given over in our unhappy times to the most portentous upheavals, find the most burning: questions of the meaning or meaninglessness of the whole of this human existence. (Husserl, 1936/1970, p. 6)

And he also warned, in Bengtsson's (1998) words, that a science that has lost its contact with the lifeworld will soon also lose its importance for everyday people. It will estrange people rather than increase their understanding or liberate them, and sooner or later this science will be called into question by researchers who discover that it has evolved into merely an instrument or an intellectual game.

Husserl's solution to the problem of a dehumanized science was to reinstate the everyday world as the foundation of science. His intent was for phenomenology, and his theory of the lifeworld, to become the new basis for all philosophy and human science research. Merleau-Ponty says in his famous preface to *Phenomenology of Perception* (1945/1995) that towards the end of his life Husserl identified *Lebenswelt* as the central theme of phenomenology. Although the notion that the lifeworld is a central theme of phenomenology is accurate, it is not entirely correct to say, as many besides Merleau-Ponty do, that this idea came late in Husserl's life. Already in the *Logical Investigations* he introduced this idea with his famous words, "Going back to the things themselves." Instead of putting scientific technicalities between the scientist and the world, he wanted science to be close to the lived world. However, he also saw the negative power in what he called "the natural attitude" of the lifeworld. The natural attitude is an uncritical and unaware position from which the world is basically understood. This attitude is a good-enough approach in everyday life, but it is all too imprecise and weak for scientific purposes. I will return to this idea later, but for now only point to the fact that the vagueness of this attitude was also one of Husserl's main motives for developing a phenomenological and transcendental philosophy.

TRANSCENDENTAL PHENOMENOLOGY

Through transcendentality Husserl aimed to go beyond the natural attitude· He offered a radical alteration of natural positing. According to this idea, we can step out of the natural attitude and "put out of action" or "exclude" or "parenthesize" parts of the world from our consciousness. Thus, a reduction has taken place. The reason for this maneuver is, as Giorgi (1997) expresses it, that we, in science and research, must problematize what we experience, instead of assuming that it is something real, in the word's most primordial meaning. Phenomenology, he says, "wants to understand what motivates a conscious creature to say that something 'is'." (p. 239) He continues,

> ... even when one encounters in experience things and events that 'obviously' have existence, the reduction directs one to step back and describe and examine them as presence.

Instead of living the world in our everyday attitude, and thus being unproblematically "present" to the things in it, Husserl wanted us to begin examining this very "presence." This reduction does not mean, however, that the "bracketed" part of the world is lost. Husserl (1913/1998) explained that even if we choose to parenthesize the world, it will always remain there for us, as an "actuality" or, more correctly, "potentiality" to consciousness. It is also important to note that Husserl never aimed at denying the existence of a real world (see Husserl, 1928/2000). Consequently, when modifying Husserl's philosophical idea for the purpose of empirical scientific analysis I found it important to introduce a new concept with which to make explicit how to change from the natural attitude to a phenomenological attitude. With the term "bridling" we can cover first of all the meaning of "bracketing" (Ashworth & Lucas, 1998; Giorgi, 1997; Husserl, 1913/1998), that is, the restraining of one's pre-understanding in the form of personal beliefs, theories, and other assumptions that otherwise would mislead the understanding of meaning and thus limit the researching openness. The term "bridling" moreover covers an understanding that not only takes care of the particular pre-understanding, but the understanding as a whole. We "bridle" the events of understanding so that we don't understand too quickly, too carelessly, or in too slovenly a manner, or in other words, so that we don't make definite what is indefinite (Dahlberg & Dahlberg, 2003).

By understanding "bracketing" as "bridling" we can better deal with the impossibility bracketing all pre-understanding; it is impossible to even grasp all pre-understanding. Being in the world and therefore being human, is a prejudice from which we can never be totally free. This is an approach to epistemology that also was adopted both by the modern hermeneutics that was developed after Husserl, by for example Gadamer, and by Merleau-Ponty, the French successor to Husserl. It is clear, that to Gadamer as well as to Merleau-Ponty we can never escape the lifeworld, the complex, qualitative and lived reality, which is there for us, whatever we do. Furthermore, Merleau-Ponty not only accepted the involvement in the world, he saw it as an opportunity. When we try to understand something in the world we must see this phenomenon in relation to other phenomena in the world. Everything that is, said Merleau-Ponty (1948/1968) is *this*, has its meanings, because its relations to everything else in the world. The world as lifeworld is there already, with its meanings and is therefore, so to speak, pre-scientific and pre-reflective. When we start thinking and doing our scientific work, we do it within the lifeworld. An important question is whether Husserl also thought that way.

The lifeworld theory puts in question any idea about a transcendental platform and makes clear that such understanding above and apart from the lived world is an illusion. The discovery of a pure consciousness was a central project for Husserl, as well as it served as "fuel" for much of his philosophical endeavors. But if this project includes solipsism it is at the same time an unreachable, unrealistic goal, especially if the lifeworld theory is coming along at the same time. Therefore, a main question is how Husserl could accomplish at all two so very different approaches to the question of how humans epistemologically relate to the world. As I refuse to throw out the baby with the bath-water and skip the whole of Husserl's phenomenology because of these complications, I want to go back to Husserl again to see what he really says about consciousness and its worldliness.

INTENTIONALITY

The main question is how humans epistemologically relate to the world. Husserl's answer to this question is that humans' relationship with the world is intentional. In addition to the themes of transcendentality and lifeworld, Husserl's theory of intentionality is central to phenomenology. In *Ideas I* (1913/1998) Husserl stated clearly that intentionality is a main theme of phenomenology because intentionality has to do with a basic way of approaching the world in which we are spontaneously engaged.

The phenomenological theory of conscious intentionality was introduced by Brentano but phenomenologically explicated by Husserl in order to epistemologically understand and conceptualize the natural attitude. Husserl's theory of intentional structures says that consciousness always is consciousness of something. Consciousness is not something by itself, but always directed toward some object, that is, the perception has its perceived, the wish its wished for, the thought its idea, etc. When we experience something, it is experienced as something that has meaning for us. There is always an intentional relationship with the things that make up our everyday lives. We understand the meaning of the things that we use and that we see around us as the things and places that belong to and signify our world.

Through intentionality, consciousness aims at presenting "whole pictures." An example of how this happens may help. When I sit in my study I can lift my gaze from the keyboard and see the yellow gable of the neighbor's house. That gable is directly given to me, or in Husserl's words, it is directly presented to my experience. But when I look at the gable, I don't think there is nothing but a gable, as a matter of fact I "see" a house. In my natural attitude, I experience a house even though the gable of the house is all that I can really see. However, even the rest of the house is given in the actual presentation and are thus "seen" by me. The back of the house is possible to confirm and to make into a direct experience by walking around the house and looking at the other side. If I do this, go out and turn around the corner and make sure that there is a house "behind" the gable, I would also see that at the end of the house is another gable, behind which there are a couple of trees. I could also open the door of the house and find that there are rooms, etc. These parts of my neighbor's place which are not immediately presented to consciousness, but nevertheless exist in my consciousness, are what Husserl (1929/1977, 1948/1973) named apperceptions or appresentations, or, the object's horizons.

These intentional structures are important to grasp if we want to understand an object's meaning in research. In the moment of perceiving we experience the world in an all-at-once way, implicitly understanding what it means. We see the particular characteristics of the object or event, whatever it may be, but more important, in the phenomenological frame of mind we can also see the aspects, or patterns, in the phenomenon that let us grasp the essence of it. An example may help again. If we see "a bottle" it means that we have understood the essence of bottle, otherwise we would say it is a glass, or a cup, or anything else. This active relationship in which we experience the things and events of our world as endowed with meaning, as *meant*, is the intentionality of which Husserl spoke.

Husserl's notion of appresentations completing the picture is comparable to the pre-structures and the implicit knowledge that Heidegger (1927/1962) and Merleau-Ponty (1945/1995) speak of, or the pre-understanding and prejudices of which Gadamer (1960/1995) speaks. Strictly, the idea of appresentation or apperception and the idea of pre-understanding are, from a philosophical perspective, not identical. For the purpose of empirical research this difference is, however, less important than the similarity.

Heidegger insisted that pre-understanding is a condition for new knowledge. Gadamer noted that pre-understanding is an unavoidable, even necessary, pre-condition for understanding and acquiring knowledge, but, as Husserl maintained about appresentations, Gadamer stated that we have to be aware of pre-understanding and suspend it, that is, question it, to prevent its negative influence on understanding. Accordingly, pre-understanding and appresentation are both intentional structures and form our pre-theoretical, non-critical, taken-for-granted knowledge about something which we acquire as residents of the lifeworld. Normally, this is not a problem as long as the natural attitude is a satisfactory approach for encountering the world. However, when the purpose is to propose or question knowledge, upon which, for example, health care, psychotherapy, and education decisions are made, the natural attitude must be exchanged in favor of a more critical attitude.

The human ability of intentionality which I have described presupposes a lifeworld. The experiences of, e.g., houses, sad and suffering people, non-learning students, etc., demand the fullness of the lifeworld. The lived experience is more capable of such an adventurous enterprise than a pure consciousness ever could be, but we must learn how to change from the natural attitude to a phenomenological attitude, i.e., a "bridled" attitude, and not make definite too quickly and negligent what is indefinite, ambiguous, and rich.

SELF REFLECTION AND SELF AWARENESS

A creative way of approaching this particular epistemological dilemma might be to emphasize the continuum, or more metaphorically speaking, the "pathway," between the extreme positions that have been mentioned. We can imagine that although the purity of the transcendental platform is impossible to reach, the path between natural attitude and the point of pure transcendence is accessible to us and provides an entry point from which we can develop an approach of critical scrutiny for research. The phenomenological attitude and "bridling" the understanding of the world and its phenomena contrasts with the natural attitude. I will point to "bridling" as an aspect of human capacity that serves as a crucial ability in human science research to think about and reflect upon our own consciousness.[1] But first, a few words about reflection in general.

Reflection is inherently paradoxical. We can never put total trust in its adequacy to disclose the world's phenomena since we tend to forget that our act of reflection, undertaken in order to discover an object's meaning, is part of the consciousness that participated in the creation of meaning in the first place. Reflection always has a blind spot that cannot be seen nor explicated, because it, as everything else, is part of the lifeworld. Reflection, as anything else, is part of the flesh of the world (Merleau-Ponty, 1948/1968). When a phenomenon presents itself to us, it is always infused with meaning. As perceiving people we are not just *receiving* this meaning but are actively contributing to it. In other words, meaning is the result of the creative relationship between the experiencing person and the phenomenon in focus. Immediately, in this initial encounter with a phenomenon, there seems to be a basic kind of reflection of which we are not aware, generally speaking. However, from an epistemological or methodological perspective, we do not refer to this first level as reflectivity. Merleau-Ponty (1945/1995, 1948/1968) explicitly pointed out that meaning does not come to be through an act of reflection.

The human lifeworld is characterized by a natural attitude, which is an uncritical and unaware position from which the world is basically understood. But within the lifeworld there is another approach, a critical one. This approach involves processes of reflecting of which we are aware. Through reflection, consciousness, which is directed towards the world, turns toward the self; consciousness establishes a distance between itself and the world, and between itself and the natural attitude in which we take for granted that the way in which we understand the world is the way that it is (Bengtsson, 1993, 1995). A reflective and distancing procedure means a move in interest from "what" to "how", i.e., *the way* that a phenomenon has meaning and knowledge is acquired becomes an object for consciousness. Merleau-Ponty's (1945/1995, p. xiii) words, "to slacken the threads of intentionality," conveys Husserl's notion of becoming distant from something without negating it. To put slack in the threads that connect us with the phenomena we perceive implies that we can strive for objectivity by examining what we believe we know about the phenomena, with the understanding that we are always part of the lifeworld and that we cannot and don't want to step outside of it in order to study it. Instead of being immersed in the natural attitude, we distance ourselves and focus more critically upon the phenomenon of interest. Distancing in this context does not mean, as it might seem, a dissociation, which leaves something behind. Rather, as Gadamer (1960/1995, p. 298) says, distancing "lets the true meaning of an object emerge fully." We recognize also the more abstract meaning of this statement when we think of how some everyday experiences sometimes seem so incomprehensible when we just "keep on living," and how they become quite simple and possible to grasp, when we all of a sudden make a break and ponder upon the difficulties, and maybe also talk about them to a friend. Sometimes putting into words is the only distance that is required. The meaning of the experience being verbalized is that the speaker or writer is distancing her or himself from the natural attitude. Paradoxically, this also means that distancing brings the human being closer to the world. Distance thus actually means a rapprochement with the object! This understanding is implied in Husserl's idea about phenomenological reduction.

However, there is a difference between the act of reflecting that looks at the thinking "I" and its processes of consciousness, and grasping the significance and the implications of doing so. We can be self reflective without being self aware. We can reflect upon our actions, behaviors, decisions and so forth, and yet remain unaware of the significance, the subjective meaningfulness that influences our actions, behaviors, and decisions.

Self-awareness is a potential capacity that can be developed, and it is at the same time something that we cannot escape if we are doing scientific research. It demands attention and explication, particularly by those of us who investigate and describe the lived experience of others. Researcher self-awareness is methodologically and ethically paramount to valid phenomenological research. Self-awareness has its origins in the lifeworld and emphasizes openness. My position here is that the gulf between those who dismiss transcendence as a goal that is impossible to achieve and those who insist that it easily can be accomplished is less problematic than it might appear. Certainly it is a tricky philosophical issue that will not be resolved here. What is important for human science research is, however, to understand that the capacity for self awareness, however limited, is part of our ontology. Since we are humans and as such a part of the world we are endowed with the ability for critical scrutiny of ourselves and the ability to seek objectivity with regard to intentionality and our own reasoning processes. It seems that we have no choice but to be always drawn toward self awareness when we undertake phenomenology, because it ultimately leads us to consider our constitutive involvement with the phenomena that we investigate and describe, it ultimately leads us to consider our belongingness to the world. It matters less that we can or cannot achieve a pure transcendence or a pure self-awareness than that we realize the importance of taking the first step on the path toward objectivity, that we develop an awareness of the conscious processes that contribute to our research and understanding.

CONCLUSIONS

In conclusion, it has been shown a way to throw out the dirty bath water and keep a vital, living baby. What remains to be explicated, then, is (1) whether or not Husserl changed his mind, and (2) whether or not his followers took up the idea of transcendentality at all.

We have to be aware that the lifeworld was one of the many ways for Husserl to reach that cherished goal of a pure consciousness that should ensure pure knowledge, which he shared with most contemporaneous philosophers. Let us, however, speculate that by that time in his life when he thematized *Lebenswelt* as a lifeworld theory, Husserl may have surrendered the dream of developing a pure transcendental phenomenology, and instead turned to the notion of the lifeworld as a median point between the difficulties with transcendence and the need for an approach with which to begin the work of phenomenology. This conclusion is supported by Husserl himself; he says:

> Everywhere the problems, the clarifying investigations, the insights of principle are *historical*. We stand within the horizon of human civilization, the one in which we ourselves now live. We are constantly, vitally conscious of this horizon, and specifically as a temporal horizon implied in our given present horizon. To the one human civilization there corresponds essentially the one cultural world as the surrounding lifeworld with its [peculiar] manner of being; this world, for every historical period and civilization, has its particular features and is precisely the tradition. (1936/1970, p. 369)

Since the lifeworld, for example in the form of tradition, constantly forces itself on us we must conclude, once and for all, that there cannot exist such a thing as a pure consciousness. The human consciousness is, so to speak, embedded in the lifeworld. So why was he not clear in this notion? It is hard to believe that Husserl did not see the obvious problems that we so clearly see now. We also know how hard it is for many scholars to give up a dream of developing something provocative new, however hopeless it seems.

On the other hand the only sufficient answer to the question of whether Husserl changed his mind is a simple "no." The transcendental idea and the lifeworld theory are simultaneous all through his philosophy. As a matter of fact, they can be understood as intertwined, presupposing each other. And, if Husserl held both ideas the whole time it is incorrect to say that he changed his mind.

This understanding of Husserl's philosophy is confirmed by some notes in *Ideas I*. For example, he asserts that phenomenologists practicing the phenomenological reduction are "not supposed to stop being natural human beings" (1913/1998, p. 149). That is, even if we pay effort into scrutinizing our own consciousness and its work, it is not the case that we let go of the lifeworld. More so this understanding is confirmed later on in *Ideas II*. For example, Husserl stated about the human-scientific attitude:

> As persons we are in relation to a common surrounding world—we are in a personal association: these belong together. We could not be persons for others if there were not over and against us a common surrounding world. The one is constituted together with the other. (1928/2000, p. 387)

Even if Husserl struggled with the creation of a pure consciousness, he at the same time seems to have been aware of the existence of certain fundamental relations between people and their world.

Husserl's investigations are more ambiguous than we usually want to accept, they lead to the disqualification of his premises. To quote Merleau-Ponty (1960/1987), we must not imagine Husserl hamstrung by vexatious obstacles, since locating obstacles was the very meaning of his inquiry. To search for disqualification of one's premises and laying out the meanings of the obstacles along the road, I would say, is the very definition of a phenomenologist!

The issues of pure description concretize the assumed split between Husserl and his followers concerning the transcendental idea. However, I argue that Merleau-Ponty and Gadamer came to the same conclusion as I have shown in this paper, that it is not the ultimate position of a pure consciousness that is of the greatest importance but the pathway between this position and that of the ambiguous lifeworld. In my reading of Merleau-Ponty and Gadamer, it has become clear that instead of abandoning Husserl they take up his discussion about how it is possible to problematize the natural attitude, even abandon it in favor of a more scientific one, and thereby "bridling" our presuppositions (Dahlberg, Drew, & Nyström, 2001). Merleau-Ponty (1948/2004) said:

> The world of perception, or in other words the world which is revealed to us by our senses and in everyday life, seems at first sight to be the one we know best of all. For we need neither to measure nor to calculate in order to gain access to this world and it would seem that we can fathom it simply by opening our eyes and getting on with our lives. Yet this is a delusion.

A few years earlier (Merleau-Ponty, 1945/1995), he had given a clear account of the natural attitude and our presuppositions:

> It is because we are through and through compounded of relationships with the world that for us the only way to become aware of the fact is to suspend the resultant activity, to refuse it our complicity (to look at it *ohne mitzumachen*, as Husserl often says), or, yet again, to put it 'out of play'. Not because we reject the certainties of common sense and a natural attitude to things—they are, on the contrary, the constant theme of philosophy—but because, being the presupposed basis of any thought, they are taken for granted, and go unnoticed, and because in order to arouse them and bring them to view, we have to suspend for a moment our recognition of them. (xiii)

> ... [The philosopher] must suspend the affirmations which are implied in the given facts of his life. But to suspend them is not to deny them and even less to deny the link which binds us to the physical, social and cultural world. It is on the contrary to *see* this link, to become conscious of it. It is 'the phenomenological reduction' alone which reveals this ceaseless and implicit affirmation, this 'setting of the world' [*thèse du monde*] which is presupposed at every moment of our thought. (p. 49)

Ashworth (1996) argues that Merleau-Ponty gave Husserl's idea of bracketing the meaning of "set[ting] aside theories, research presuppositions, ready-made interpretations, etc., in order to reveal engaged, lived experience" (p. 9). Bracketing, or "bridling" as I prefer to see the idea, refers to the adoption of an open researching attitude to the world as it presents itself, the most central idea of phenomenology. At the same time as this interpretation of Husserl makes the reading of his philosophy worthwhile it makes clear that this understanding of the phenomenological reduction can involve a transcendental ego.

Aware that we are anchored to a world, Merleau-Ponty did not try to go beyond it. He actually never had that aim. For him a reflective stance did not withdraw from the world. On the contrary, he emphasized a phenomenological reduction as a way to come closer to the world. Reflection, Merleau-Ponty (1945/1995) said,

> steps back to watch the forms of transcendence fly up like sparks from a fire; it slackens the intentional threads which attach us to the world and thus bring them to our notice; it alone is consciousness of the world because it reveals that world as strange as paradoxical. (xiii)

Scientific openness to the phenomenon in focus was of crucial importance also for Gadamer. He laid out for us a process of hermeneutics that is firmly rooted in phenomenology, and acknowledged the difficulty for scientists that pre-understanding presents: 'Understanding begins . . . when something addresses us. This is the first condition of hermeneutics. We know now what this requires, namely the fundamental suspension of our own prejudices' (Gadamer, 1960/1995, p. 299). He further explained this idea:

> The fundamental elimination of prejudices that science requires of its researchers may well be a laborious process, but it is always easier than overcoming the illusions that constantly arise from one's own ego (that of an individual, group, people, or culture to which the person belongs and listens) in order to see what is. (Gadamer, 1983/1998, p. 31)

A "bridled" and reflecting attitude means a balancing act where we want to keep a scientific openness at the same time as we are aware of our worldliness. As Gadamer (1960/1995) put it, the meaning of this balancing endeavor means that we have to "distinguish the true prejudices, by which we *understand*, from the *false* ones, by which we *misunderstand*" (p. 299). This is a tricky business. It is hard to "separate in advance the productive prejudices that enable understanding from the prejudices that hinder it and lead to misunderstandings" (ibid, p. 295).

Thereby I want to conclude that Merleau-Ponty's and Gadamer's epistemological analyses are not that different from Husserl's after all. I have given a few examples of texts that show how these two followers of Husserl take seriously upon his hard work of understanding human consciousness. They also show that Merleau-Ponty and Gadamer elaborate the idea of a scientific attitude that is fundamentally different from the natural attitude.

I want to end this chapter by remarking that phenomenology by no means can be understood as one single idea; on the contrary, paraphrasing Palmer about hermeneutics I would like to state that there are phenomenologies. However, these phenomenologies cannot be seen as discordant. They have to be understood as harmonious strivings to develop a whole new scientific approach to the human world, an approach that acknowledges and embraces complexity without doing harm. The connecting thought within these phenomenologies is composed of the lifeworld, a reflective intentionality and understanding self awareness, ideas that Husserl developed because of his infinite struggle to explicit a transcendental ego.

NOTE

1. Both Natanson and Fink, well-known interpreters of Husserl's philosophy, encourage us to propose that when Husserl discusses transcendentality he not only has a pure consciousness in mind, but also implies self-reflection and self-awareness as something close to this ultimate position.

REFERENCES

Ashworth, P. (1996). Presuppose nothing! The suspension of assumptions in phenomenological psychological methodology. *Journal of Phenomenological Psychology, 27*(1), 1–25.

Ashworth, P., & Lucas, U. (1998). What is the "World" of Phenomenography? *Scandinavian Journal of Educational Research, 42*(4), 415–431.

Bengtsson, J. (1993). Theory and Practice: Two fundamental categories in the philosophy of teacher education. *Educational Review, 45*(3), 205–211.

Bengtsson, J. (1995). What is reflection? On reflection in the teaching profession and teacher education. *Teachers and Teaching: Theory and Practice, 1*(1), 23—32.

Bengtsson, J. (1998). *Fenomenologiska utflykter* (Phenomenological excursions). Gothenburg, Daidalos.

Dahlberg, H., & Dahlberg, K. (2003). To not make definite what is indefinite. A phenomenological analysis of perception and its epistemological consequences. *The Humanistic Psychologist, 31*(4), 34–50.

Dahlberg, K., Drew, N., & Nyström, M. (2001). *Reflective lifeworld research*. Lund: Studentlitteratur.

Gadamer, H.-G. (1995). *Truth and method* (2nd rev. ed.) (J. Weinsheimer & D. Marshall, Trans.). New York: The Continuum Publishing Company. (Original work published 1960)

Gadamer, H.-G. (1998). *Praise of theory. Speeches and essays* (C. Dawson, Trans.). New Haven: Yale University Press. (Original work published 1983)

Giorgi, A. (1997). The theory, practice, and evaluation of the phenomenological method as a qualitative research procedure. *Journal of Phenomenological Psychology, 28*(2), 235—260.

Heidegger, M. (1962). *Being and time* (J. Macquarrie & E. Robinson, Trans.). Oxford: Blackwells. (Original work published 1927)

Husserl, E. (1970). *The crisis of European sciences and transcendental phenomenology* (D. Carr, Trans.). Evanston, IL: North Western University Press. (Original work published 1936)

Husserl, E. (1973). *Experience and judgement* (J.S. Churchill & K Ameriks, Trans.). Evanston, IL: North Western University Press. (Original work published 1948)

Husserl, E. (1977). *Cartesian meditations* (D. Cairns, Trans.). The Hague: Martinus Nijhoff. (Original work published 1929)

Husserl, E. (1998). *Ideas pertaining to a pure phenomenology and to a phenomenological philosophy. First book.* (F. Kersten, Trans.). Dordrecht: Kluwer Academic Publisher. (Original work published 1913)

Husserl, E. (2000). *Ideas pertaining to a pure phenomenology and to a phenomenological philosophy. Second book.* (R Rojcewicz & A. Schuwer, Trans.). Dordrecht: Kluwer Academic Publisher. (Original work published 1928)

Merleau-Ponty, M. (1968). *The visible and the invisible* (A. Lingis, Trans.). Evanston, IL: Northwestern University Press. (Original work published 1964)

Merleau-Ponty, M. (1987). *Signs* (R. McCleary, Trans.). Evanston, IL: North Western University Press. (Original work published 1960)

Merleau-Ponty, M. (1995). *Phenomenology of perception* (C. Smith, Trans.). London: Routledge. (Original work published 1945)

Merleau-Ponty, M. (2004). *The world of perception* (O. Davies, Trans.). London: Routledge. (Original work published 1948)

CHAPTER 6

HUSSERL AGAINST HEIDEGGER AGAINST HUSSERL

PAUL S. MACDONALD

If *Being and Time* is written "against" someone, then [it is] against Husserl.
(Heidegger, Letter to Jaspers, 1926)

Everything here [in *Being and Time*] is a translation and transposition of my thought.
(Husserl, marginal note, 1928)

Against Heidegger... an original motive lies, for science as for art, in the necessity of the game.
(Husserl, unpublished manuscript, 1931)

I wrote a whole book under [Husserl's] inspiration: *L'Imaginaire*. Against him, granted—but just insofar as a disciple can write against his master.
(Sartre, War Diaries 1939–1940).

INTRODUCTION

So many ways to say that one is *against* someone, especially in the context of also saying that one agrees with or follows someone (as we shall see below). How far does one have to be no longer *with* a thinker's thought to be *against* that thinker's thought? Heidegger, Jaspers, Sartre, and Merleau-Ponty repeatedly acknowledged in print that one of their points of departure was in Husserl's phenomenological approach. But equal to and parallel with their claims to be going along with Husserl were their vigorous declamations of his mistakes, dead-ends and failures. It seems to have been vitally important to thinkers during the 1920s and 1930s to create as much distance as possible between themselves and their former "master." Less well remarked perhaps is the discrepancy between the Existentialists' complaints about Husserl's aborted achievements and what Husserl actually delivered. Husserl's protracted reflections on any given issue are peculiarly difficult to expound, and collaterally it is often very hard to identify a specific doctrine or discursive position in such a way that it is clear *what* assertion his critic is attempting to rebut. At any given time, Husserl the mediator was rarely satisfied with his written work; he thought of his analytic studies as "a process of endless corrections and revisions." The few works that he published in his lifetime look like "purely momentary states of rest, or 'condensations' of a thought movement that was constantly in flux." (Bernet, Kern, & Marbach, 1992, p. 2) In 1925, Heidegger himself said that "it is characteristic of Husserl that his questioning is still fully in flux, so that we must in the final analysis be cautious in our critique." (Heidegger, 1982b, p. 121) Husserl's thought on central phenomenological notions never stood still; he was, in his own memorable image, "an endless beginner," and would be the first to open a new path when the woods became lost in the trees. Heidegger's constant early references to pathmarks, pathways, and so forth (see esp., van Buren, 1994, pp. 5–9) are preceded by Husserl's own favorite imagery of phenomenology as a series of paths, a journey, or a voyage. (see, MacDonald, 2000, pp. 82–85)

Some of the Existentialists' specific criticisms of Husserl's assumptions and approach are as follows. First, for Husserl, the dominant dimension of intentional directedness is cognitive, that is, phenomenological analysis of the total unity of intentional act-and-object is calculative and componential. Second, he accords the greatest epistemic weight for human understanding on rational insight and higher-order intuition, specifically the ability to *reflect* on one's experiences. Third, the transcendental ego as uncovered through the reduction is a detached spectator, unengaged with the "natural" world and its everyday things and values. Fourth, consciousness is conceived as a monadic unity through which, or into which, nothing can penetrate; this conception leads to an irretrievable solipsism. Fifth, the final stage of the reduction is meant to bring about an exact science of essences, i.e., material and spiritual essences, correlative to the categories of a formal ontology; but one thing, human being, resists such categorization, since it does not *have* an essence. Sixth, and perhaps the most damning, the methodical technique of phenomenological epoché, required in order to bracket the world's being, cannot be fully performed, that is, not without seriously damaging the actual relation that conscious beings have with their world.

This capsulized critique, however, does not take adequate account of Husserl's own answers to some of these charges; though, of course, like any thought-capsule it's handy for didactic purposes. Over and over again, for more than 12 years, between 1912 and 1925, Husserl worked on clean copies of the Second and Third Books of the *Ideas*. One after the other, his personal assistants prepared new drafts, only to have them returned later revised and corrected; they despaired that the other parts of his great work would never reach publication. But Heidegger read the manuscript version two years before the publication of his own *Being and Time* in 1927 (Heidegger, 1982b, p. 121; 1962, p. 469, note ii), and Merleau-Ponty read them, in "a near rhapsody of excitement" just before the outbreak of war in 1939. (Merleau-Ponty, 1996, App.) Many of the charges leveled against Husserl have responses in these texts, and more than just responses, they *further* Husserl's project toward an emergent existential and inter-subjective phenomenology.

Husserl's constant efforts to revise and correct his position become evident, for example, in the way in which these texts deal with the first and second charges above. What Husserl in *The Crisis of European Sciences* (1936) calls the lifeworld (*Lebenswelt*) in *Ideas Book Two* he calls the surround-world (*umwelt*), an idea conveyed very nicely by the word *environ*-ment. (See D. W. Smith, in Smith & Smith, 1995, pp. 360–362; J. C. Evans, in Nenon & Embree, 1996, pp. 57–65) The *umwelt* is one's immediate environment; the world which surrounds an animate conscious being, already structured in determinate ways; that is, things and persons and events already have value *before* predicative judgments are made about them. "The surround-world is in a certain way always in the process of becoming.... To begin with, the world is, in its core, a world appearing to the senses and characterized as *on hand*, a world given in straightforward empirical intuitions and perhaps grasped actively." (Husserl, 1989, p. 196) This low level, non-reflective engagement with handy things, items that fit in one's hand, is not gauged in mere use-terms, it also uncovers the direct grasp with value-terms. Husserl continues: "The ego then finds itself related to this empirical world in new acts, e.g., in acts of valuing or in acts of pleasure and displeasure.... There is built upon the substratum of mere intuitive representing an evaluating which (if we presuppose it) plays in the immediacy of its lively motivation, the role of value-perception... in which the value character itself is given in original intuition." (Husserl, 1989, p. 196)

Husserl's comments on the cognitive grasp of use-values embedded in ordinary worldly things is strikingly similar to Heidegger's better known exposition of the different "kinds" of items found in one's environment. Such ordinary mundane items are either *available*, "at" hand (*zuhanden*), or they are *occurent*, "for" hand (*vorhanden*); for example, the hammer grasped as an item imbued with use or service, in contrast with an "inert" piece of metal or wood. Husserl's notion of originary value-perception is hardly surprising since in the same section he goes on to illustrate the notion of an object "apprehendable as in the service of the satisfaction of such needs" as hunger and warmth with "heating materials, choppers, hammers, etc." In his 1925 lectures, he devotes considerable attention to the kind of intentional behavior that deals with psychically endowed objects such as artifacts. (Husserl, 1977, pp. 84–88) He says that such items-on-hand are directly grasped as use-objects, in contrast to those that do not have use-value, objects just lying about. This grasping or apprehending is not an operation of explicit awareness, no judgment is formed by means of which one could *infer* or derive value, it is thus a *pre*-predicative awareness. Useful items are associated in the surround-world through "a web of intentions," interlinked by way of meaningful indications, much in the same fashion that Heidegger describes the environing world of the ready-to-hand as "a referential totality." One's experience of the manifold connections of all use-objects, artifacts and cultural products, says Husserl, is an experience of *motivated* relations.

Long before the earliest draft of the sections of *Being and Time* devoted to this topic Husserl had clearly underlined the need to make explicit, or bring to the fore, the pretheoretical "meaning" attached to or embedded in useful versus inert objects. In his Lectures on Nature and Spirit from 1919, he states, "I understand the significational unity that the word 'hammer' expresses by relating it back to that which posits the end, to the subject creating at any time useful means for purposeful productions of a definite type." The practical and bodily grasp of useful meanings, i.e., items endowed with meaning *for use*, is built on an essential understanding of motivational relations which extend through and beyond those particular items. "All significance objectivities and significance predicates are judged in their manner of correlation as rational and irrational. [In this sense of 'rational'] the hammer is 'to be useful,' but it can be a good or a bad hammer.... The 'is to be' expresses the pretention [forward-directed intention], it expresses that it stands under the ideas of reason." (Husserl, in Nenon & Embree, 1996, pp. 9, 10) In his exposition of this and associated texts, Ulrich Melle quotes from another manuscript: in our everyday experience we apperceive our environment so immediately with spiritual predicates of meaning, "that these predicates are downright designated as perceived, as seen, as heard, etc., just as the real predicates which are given in the most immediate sensuous experience." (Nenon & Embree, 1996, p. 24) In *Ideas Second Book*, Husserl again made an attempt to account for human practical understanding of use-endowed items in the world of their immediate concern. The use which an item has for its user motivates (not causes) one to take it up, as well as inspiring other motivations tied in with other use-objects. "They now engage [the ego's] interest in their being and attributes, in their beauty, agreeableness, and usefulness; they stimulate its desire to delight in them, play with them, use them as a means, transform them according to its purposes, etc.... In a very broad sense, we can also denote the personal or motivational attitude as the *practical attitude*." (Husserl, 1989, p. 199)

The question of reciprocal influence between the two thinkers during those years must take account of a complex plot-line. In his early academic career, Heidegger cited Husserl's *Sixth Logical Investigation* as the best and most important point of departure for his ontological investigations. As a university student Heidegger read Husserl's *Logical Investigations* with great diligence; in fact, he kept the library copy in his room for two years since no one else seemed to want it. Fifty years later he recalled the spell which Husserl's early work exerted on him: "I remained so fascinated by Husserl's work that I read in it again and again in the years to follow.... The spell emanating from the work extended to the outer appearance of the sentence structure and the title page." (Heidegger, 1998, p. 81) Near the end of his life, in his 1973 seminar on his way to phenomenology, he remarked that the essential discovery and "burning point" of Husserl's thought was the notion of categorial intuition of being, which then became an "essential motivating force" for his own thought. He said that, in the Sixth Investigation, Husserl "brushed against the question of being" (quoted in van Buren, 1994, p. 204); and that "Husserl's accomplishment consisted precisely in this making present of being that is phenomenally present in the category. Through this accomplishment . . . I finally had a basis—'being' is no mere concept, is no pure abstraction, which arises in the course of a derivation." In his lecture courses in Winter 1921–1922, Summer 1922, Summer 1923, Winter 1923–1924, and Summer 1925 (amongst others), Heidegger devoted considerable attention to the *Logical Investigations* (van Buren, 1994, p. 205).

As early as his 1919 lecture course Heidegger insisted that a fundamental destruction (i.e., "deconstruction") of phenomenological inquiry was needed. His critique attempted to trace the founded, theoretical articulation of such Husserlian concepts as intentionality, content, relation, enactment, time, sense, being, and truth back to the genuine phenomenological stratum (*life in and for itself.*) These central Husserlian concepts needed to be rethought from their source in a pretheoretical understanding of the world. He conceived of his own historical ontology as "the most radical phenomenology that begins in the genuine sense 'from below'," i.e., in the depths of human practice. It was to be a new genuine beginning, now understood not merely "in its actuality as a philosophical movement," but as "a possibility of showing the things themselves." (van Buren, 1994, p. 205) Husserl's notion of thing-hood was rethought in terms of the significance and lived spatiality of the lifeworld, the everyday world experienced prior to theoretical insight. Van Buren says that, although Heidegger borrowed the very idea of lifeworld from Husserl, "he took the practical sense immediately encountered here to be the primal stratum of lived experience."—but that is exactly the import of Husserl's analysis of preobjective, pretheoretical experience.

In order to augment Husserl's phenomenological analysis, Heidegger brought Dilthey into his rethinking of an ontological inquiry and said that "we are indebted to him for valuable intuitions about the idea of this science," psychology as a descriptive science. According to van Buren, "The secret longing of his life began to be fulfilled by phenomenology. Heidegger wanted to follow Dilthey's lead in turning toward a basic science of human life as it is lived, and not as it is theorized about. He said that Dilthey was the first to understand the aims of phenomenology . . . the essential point here is not so much the conceptual penetration as the sheer disclosure of new horizons for the question of the being of acts and, in the broadest sense, the being of human being." He also claimed that Dilthey had opened up three existential horizons: "world" (content sense), the "total person" (relational sense), and "historical context" (temporal sense). Heidegger thought that Dilthey had moved Husserl's investigations into the lived experience of the practical and cultural world: "the person in his particular selfhood finds himself over against a world upon which he acts and which reacts upon him"; and further, that "this whole context—self and world—is there in each moment." (van Buren, 1994, pp. 208–209) Using Dilthey's hermeneutic theory he transformed Husserl's theoretical method of universal intuition into his own existential hermeneutics, an understanding of how human beings comport themselves in an understanding manner. "The phenomenological criterion," he wrote in 1919, "is solely the understanding evidence and the evidential understanding of lived experiences, of life in and for itself as its *eidos*." But he denied Husserl's claim that philosophical inquiry could arrive at any kind of transtemporal essences, to which an exact phenomenological terminology would ideally correspond. The traditional philosophical concepts were no more than formal indications that provisionally orient the search for the historical "facts" about being. "Husserl's recovery of Greek ontology was thus to be pushed in the direction of Heidegger's own pretheoretical 'science of the primal source', i.e., 'a phenomenological (existential, historical-cultural) 'ontology'." (van Buren, 1994, p. 211)

In his early review of Jaspers' *The Philosophy of Worldviews* (1919), Heidegger remarked that "in the first breakthrough of phenomenology with its specific goal of originally appropriating anew the phenomena of *theoretical* experiencing and knowing ... the goal of research was the winning of an unspoiled seeing of the sense of the objects experienced in such theoretical experiencing, i.e., the how of its being experienced.... But the possibility of the radical understanding and genuine appropriation of phenomenological tendencies rests on the fact that ... experiencing is seen in its authentically factical context of enactment in the historically existing self." (Cited in van Buren, 1994, p. 209) It is not suitable, he went on, for the other spheres of existence (in Kierkegaard's phrase), the aesthetic, the ethical and the religious, to be studied in an analogous manner to the theoretical region, since this would make them over into theoretical domains. Instead, the "concrete self" revealed in these other spheres is "to be taken up into the starting point of problems and brought to givenness as the authentic, basic stratum of phenomenological interpretation, namely, one that remains related to the factical experience of life as such." Heidegger's proposed destruction of western metaphysics sought to demythologize Husserl's theoretical concept of being by dismantling it back to and reinscribing it from out of the primal stratum of content-sense in the lifeworld and environing world (*umwelt*). He wrote to Jaspers in 1926 that if *Being and Time* is "written 'against' someone, then [it is] against Husserl," and his own marginal comments in the text of sections 18–21 on Descartes' view of objects as present-at-hand indicate that he thought of this as "a critique against Husserl's stratified construction (*Aufbau*) of 'ontologies'." (van Buren, 1994, p. 210)

In his Kassel Lectures of 1925, Heidegger maintained that "in its first breakthrough, phenomenological research limited itself essentially to theoretical experience, to thinking.... Husserl misunderstood his own work." (van Buren, 1994, p. 209) Even when Husserl deals with affect, perception, and practice he models these kinds of intentionality on the attitude of "mere being-directed-toward mere objects" that characterizes theoretical thinking by way of its associated modes, i.e., "intuition," "knowing," "judgment," and "assertion." Heidegger detected an overtheorization effect in Husserl's analysis of human comportment: he alleged that, for Husserl, "every directing-itself-toward (fear, hope, and love) has the feature of directing-itself-toward which Husserl calls *noesis*. In as much as *noein* is taken from the sphere of theoretical knowing, any exposition of the practical here is drawn from the theoretical." For Husserl, in *Ideas, First Book*, the term *noesis* signifies an intentional act's distinctive cognitive mode, whereas *noema* signifies an intentional object's ideal content. Thus, on Heidegger's view, Husserl failed to understand that "the so-called logical comportments of thinking or objective theoretical knowing represent only a particular and narrow sphere within the domain of intentionality." Husserl's most fundamental concept of intentionality as "mere being-directed-toward" is a version of the theoretical attitude and thus "delives" (takes the life out of) lived experience (*Erlebnis*). Intentionality is an ultimate desideratum for Husserl, but how does one explain its ultimate status? "What has always disturbed me," he confessed, "is the question, did intentionality fall from heaven? If [it is] something ultimate [then] in which ultimacy is it to be taken? [It is] certainly not secured in a specifically theoretical discovery and experience.... Accordingly, intentionality is the formal and basic structure for all categorial structures of facticity." Against this tendency to reach an ultimate, Heidegger set forth his own existential structure of care, since "caring is the basic sense of the relation of life"; there is "the complete sense of *intentionality* in what is original." (van Buren, 1994, pp. 213, 215)

At the same time, Heidegger extolled Dilthey's efforts to conserve the "life"—sense of human lived experience, since Dilthey "wants to get at the totality of the subject which experiences the world and not to a bloodless thinking thing which merely intends and theoretically thinks the world." No one person is mentioned by name in his sarcastic reference to the bloodless, lifeless ego, but there can hardly be any doubt that Heidegger himself, as well as his listeners, would have thought Husserl to blame for this construction. In Heidegger's thoughts the bloodletting seems to have occurred through the purge of the reduction, since Husserl's reduction leads back to "an idealized absolute subject," one who is not human, but rather godlike. He poses the rhetorical question whether or not the pure ego is "an old myth" about "a fantastically idealized subject," in other words, "a fantasy of life and thinking," the hyperbolic "pure ghost," the "residue of Christian theology". (van Buren, 1994, p. 214)

Heidegger claimed that Husserl, like Jaspers, began his study of human life within an aesthetic ground-experience; "this means that the authentic relational sense of the primary experience that puts forth the object 'life' in advance is an intuiting, a contemplating of something." The eidetic reduction to the eidos-ego (the essence of any ego) involves "giving up the ground upon which alone the question of the being of the intentional could be based, namely, the factical life of Dasein." Husserl thus also failed to raise the question of the *being* of human being, that being for which intentionality is the unique structure. Having begun his quest with an approval of the Cartesian *point d'appui*, Husserl became stuck fast with an imponderable *sum* for the theoretically defined *cogito*. Husserl's methodical center is the isolated ego, *solus ipse*, and, despite repeated efforts, he is unable to extricate his thought from this complete isolation. This severe and ineradicable limitation prevents Husserl from recognizing the pre-givenness of other egos in the surround world, the historical embeddedness of every human being, and the pre-objective features of objects as available for humans' bodily comportment.

Heidegger admits in the lectures on the *History of the Concept of Time* that his critical remarks are based on Husserl's *published* works, i.e., the *Logical Investigations, Ideas First Book*, and *Philosophy as Rigorous Science*. He says that he is aware that Husserl has gone beyond the position that he is about to present here and that Husserl is aware of his criticisms, but "essentially makes allowances for them, so that my critique today no longer applies in its full trenchancy." He quotes from Husserl's letter about his own recent lectures on nature and spirit and the contents of *Ideas Second Book*: "accordingly," he comments, "the account which was first presented here is in some ways antiquated." (Heidegger, 1982b, p. 121) But not, on his view, out of date enough to abandon the account; perhaps it is symptomatic of philosophy lecturers that it is very hard to give up good arguments, even when one has good reason to think they are flawed. But many of Heidegger's most serious objections to Husserl's phenomenology were too important to his own forward movement to accommodate counter-responses. One of his main points of issue concerns the "actual theme" of phenomenology, that is, "the definition of the being of consciousness with regard to the way it is given in the natural attitude. This primary kind of experience . . . turns out to be a theoretical kind of experience and not a genuinely natural one, in which what is experienced could give itself in its original sense. Instead, the manner in which what is experienced gives itself here is defined by the feature of an objectivity for a theoretical consideration of nature, and nothing else." This statement condenses a crucial equivocation in the use of the word "natural": on one hand, "natural" in the sense of ordinary, taken-for-granted, and unreflected; on the other hand, "natural" in the sense of being oriented toward or conforming with the world of nature. Husserl carefully distinguishes between the *natural* attitude, i.e., the ordinary way in which humans comport themselves toward the world *before* reflection steps in, and the *naturalistic* attitude, i.e., a specific theoretical orientation towards objects that considers them as "parts" of the world, conformable to physical laws of motion and change. (Husserl, 1989, pp. 147–150) (On this topic see esp. Nenon, in Nenon & Embree, 1996, pp. 223–235) But Heidegger quickly draws an inference from his previous assertion: "it thus follows that the starting point for the elaboration of pure consciousness is a *theoretical* one. At first, naturally [*sic*], this in itself would not be an objection or a misfortune, but surely it is afterwards, when, on the basis of the pure consciousness derived from this *theoretical* basis, it is claimed that the entire field of comportments may also be determined, especially the practical." (Heidegger, 1982b, p. 117)

Unfortunately, however, these criticisms outran any response from the elderly Husserl and made an impact on many younger thinkers who attended Heidegger's lectures. One exemplary case study is the influence they made on Emmanuel Levinas who, in his first published work in 1930, *The Theory of Intuition in Husserl's Phenomenology*, preserves and promotes the over-theorization critique of Husserl's notion of intentional directedness. When Levinas presents Husserl's concept of intuition of essences (*wesenschau*) he claims that Husserl argues that natural scientific theory requires idealized aspects of objects, whereas in our concrete lifeworld perceptual objects appear with inexact essences. "Even though [Husserl] attains the profound idea that, in the ontological order, the world of science is posterior to and depends on the vague and concrete world of perception, he may have been wrong in seeing the concrete world as a world of objects that are primarily perceived. Is our main attitude toward reality that of theoretical contemplation?" (Levinas, 1973, p. 118) Levinas agrees with Heidegger's assessment of Husserl's failing and his recommendation for the proper theme; "Is not the world presented in its very being as a center of action, as a field of activity or of *care*?" (Levinas, 1973, p. 119) Heidegger's analysis of care as the basic comportment of Dasein towards its own and other's concerns means that Dasein has an *interest* in those things that concern it; though he cautions against excessively distractive interests which can lead to pointless curiosity, fascination and absorption in the everyday. (Heidegger, 1962, pp. 214–217) Surely, it is characteristic of the theoretical attitude that it is basically *disinterested* in the "object" of its reflection. Heidegger's *apercu* that the word "interest" is derived from *inter-esse*, "to be in the midst of," indicates that "disinterest" would mean to "step out of" this "being in the midst of." Levinas feels content to make the charge of intellectualism against Husserl in this respect; for Husserl, "the primary and fundamental attitude when facing reality is a pure, disinterested contemplation which considers things as 'merely things'. Value predicates or the characters that make a thing useful *qua* making it useful come only later. The world of theory comes first." (Levinas, 1973, p. 128; see also p. 94)

Levinas is aware of passages in *Ideas First Book* that draw attention to the intentional realm of values and uses, and quotes from an idiomatic text. In an analogous fashion to the modes of belief (the doxic thesis) about objects, new senses (meanings) of objects can be apprehended that are *founded* on the *noesis* underlying the grasp of value or use. These new senses are not "parts" of a mere thing, the way its color or shape are parts, "but instead [are] *values of things*, value-qualities, or concrete objects with values: beauty and ugliness, goodness and badness; the use-object, the art work, the machine, the book, the action, the deed, and so forth." (Husserl, 1982, p. 277) Levinas goes on to state that the meaning of existence is different in each category of "object," i.e., sensible object, value-object and use-object: "a theoretical position which Husserl calls a *doxic thesis* is always included in the act which posits these different objects as existing.... This doxic thesis is the element of intentionality which, according to Husserl, thinks of objects as existing. It is *because* each act of consciousness includes a doxic thesis that the objects of these acts—values, useful objects, or aesthetic objects—exist." (Levinas, 1973, p. 134) But this is *not* true of the text that Levinas cites and it mistakes the concept of doxic thesis. Husserl thinks of intentional position-taking as the holding of an "object" in an intentional act, a cognitive attitude so primitive that it never occurs "on its own" (so to speak) except in the form of an empty intending; the doxic thesis is always modalized in one form or another, e.g., as possible, likely, dubious, etc., and this is done through affirming, denying, and so forth. (Husserl, 1982, p. 267) When cognition turns toward this primal positing the doxic thesis is transformed into being-possible, being-likely, being-necessary, etc.,—thus, it is not true that every positing takes its object as existing. But Levinas is confident enough of his reading to indict Husserl yet again; "Husserl's assertion here demonstrates that the notion of existence remains for him tightly bound to the notion of theory, to the notion of knowledge, despite all the elements in his system which seem to lead us to a richer notion of existence than mere presence of an object to a contemplative consciousness." (Levinas, 1973 p. 134)

FIVE FOCI OF CRITIQUE

In order to sharpen the focus of this debate, let us try to tease out the various Heidegger-inspired criticisms of Husserl's approach and weigh them against Husserl's own words. There are five principal criticisms:

1. Over-theoretization (theoreticism)
2. Over-intellectualization (intellectualism)
3. Splitting of the ego
4. Consciousness and world separated by an abyss
5. Neglect of the being of the intentional

Overtheoretization (Theoreticism)

(a) All forms of human comportment toward the world are construed as analogues to a theoretical attitude, one that observes from a detached perspective; (b) the genetically primitive attitude, i.e., first in the factual constitution of the human individual, is one of theoretical interest in and observation of the world; (c) the theoretical attitude of the natural sciences is imported back into the pretheoretical understanding of the lifeworld.

Our discussion (above) of the importance of practice and value in *Ideas Second Book* for Husserl's analysis of human comportment in the surround world, and the errors in Levinas' construal of Husserl's position, seriously undermine this criticism. Dermot Moran wonders whether Husserl actually overstressed the cognitive dimension in human experience, at the expense of the practical dimension of human existence. He admits that Husserl *did* focus more on elucidating cognitive acts rather than on emotions or actions, "but in no sense did he downgrade the practical and emotive in relation to our specifically cognitive achievements." (Moran, 2000, p. 60) Due to the great difficulty in delineating the forms of fulfillment of non-cognitive experience, Husserl devoted little attention to these themes (as Hart also says, 1990, p. 211). However, Moran points out, in *Formal and Transcendental Logic* Husserl "insists on the importance of appreciating the practical forms of fulfillment with which we are most familiar, and regrets that philosophers (including his earlier self) have become preoccupied with fulfillment in theoretical disciplines such as mathematics." Moran also speculates that Husserl's Lectures on Thing and Space in 1907 are "very close to and may indeed have partially inspired Heidegger's account of the practical intentionality involved in our everyday absorption in the world." (Moran, 2000, p. 61) Husserl was fascinated with the way in which one attitude can give way to another attitude; his texts and lectures are peppered with references to "transition," "change into," "stepping forth," "stepping back," and so forth.

Heidegger actually agrees with Husserl that the theoretical, disinterested attitude is *just one* possible orientation towards worldly things that humans can adopt. Moran quotes from an unpublished Husserl manuscript of 1931 labeled "Against Heidegger" (see especially the Introduction to Husserl, 1997), where he says that the theoretical interest is motivated, like the artistic interest, by a desire to play freely, away from everyday concerns; this theoretical curiosity is by no means a deficient mode of the practical as Heidegger insists. "Special motives are required in order to make the theoretical attitude possible and, against Heidegger, it does appear to me that an original motive lies, for science as for art, in the necessity of the game (*Spiel*) and especially in the motivation for a playful 'intellectual curiosity', one that is not springing from any necessity of life, or from calling, or from the context of the goal of self-preservation, a curiosity which looks at things, and wants to know things, with which it has nothing to do. And no 'deficient' praxis is at stake here." (Moran, 2000, p. 62) When Heidegger bypasses Husserl's complex discussion of the kinds of intentional acts and their fulfillment in order to concentrate on the question of being, Moran concludes, he "succumbs in fact to a philosophical error which he is quick to diagnose in others in the philosophical tradition, namely, *leveling off* the achievement." Where Husserl's attention to the motivated theoretical perspective as emerging from the practical attitude offers more positive scope to the special character of scientific knowledge, Heidegger views it as a falling away from Dasein's ownmost authentic being. "In this respect, Husserl's interest in intentionality for the theory of scientific knowledge may have longer currency than Heidegger's attempts to transform the problematic of intentionality into the question of being." (Moran, 2000, p. 63)

Husserl's endorsement of the importance of practice, moreover the *background* of shared practices (much discussed recently by Hubert Dreyfus and John Searle) leads us to an answer to one of the other charges against Husserl, his alleged emphasis on theory at the expense of practice, and of reflection to the detriment of lived action. (see Barry Smith, in Smith & Smith, 1995, pp. 413–415) But to denigrate Husserl's supposed reliance on theoretical insight in his elaboration of the intentional structures of consciousness is to ignore his explicit subordination of theory in the transition to consideration of the human being in its psychical constitution. "What we are seeking does not lie in the consequences of theoretical, mediate thinking but in its beginnings; we are looking for its most originary presuppositions .... Legitimate theory cannot accomplish anything other than the predicative determination, in mediate thinking, of that which was first posited by originary presenting intuition (in our case experience) .... There what the 'analysis of origin' has drawn from originary intuition as the originary sense of the object cannot be annulled by any theory. It is the norm which must be presupposed and to which all possible theoretical cognition is rationally bound." (Husserl, 1989, pp. 96–97) To adopt a theoretical or detached attitude towards pretheoretical, prereflective consciousness is like attempting to compel an umpire or referee to perform *his* function while acting like a player on the field. Husserl himself was aware of the possible slippage between a theoretical attitude *about* the origins and structures of the lifeworld into a belief that a theoretical attitude can be found *within* the lifeworld. The danger here is that one might be tempted to unwittingly insert backward into one's lifeworld analysis an observer's point-of-view which is then "discovered" to be in place *after* reflection. Perhaps the Cartesian meditative ego, contingently connected to its body, could become detached enough from its surroundings, and could indeed "float above this world, above this life," but Husserl's human being is intimately bound together with its body.

However, it is *another* serious question whether or not Heidegger's exposition of the meaning of being as the preeminent phenomenon of human being-in-the-world can be *reduced to* or entirely explained in terms of the background of social practices, as Hubert Dreyfus has attempted in *Being-in-the-world.* Dreyfus' highly influential interpretation of Heidegger's criticisms of Husserl's concepts of intentionality, the detached spectator, the self-constituting subject, and so forth, grossly fail to take account of other relevant Husserlian texts on the issue of practice. Føllesdal reported that after Husserl came to Freiburg in 1916, "he clearly became more and more aware that our practical activity is an important part of our relation to the world.... There is, according to Husserl, 'an infinite chain of goals, aims, and tasks' that our actions and their products relate to." Føllesdal tried to figure out who deserved the credit for first grasping the phenomenological significance of human understanding via the circumspect background of practical activities, about which he states, "Husserl had ideas similar to those of Heidegger long before *Being and Time* was published. These ideas started appearing in Husserl shortly after he arrived in Freiburg and met Heidegger in 1916. It is possible that Husserl influenced Heidegger in this 'practical' direction.... However, it is also possible that it was Husserl who was influenced in this direction through his discussion with the younger Heidegger." (Føllesdal, 1979, pp. 372, 376) Dreyfus quotes Mark Okrent in his footnotes to the effect that "As soon as one realizes that, for Heidegger, intentionality is always practical rather than cognitive and that the primary form of intending is doing something for a purpose rather than being conscious of something, the structural analogies between the argument strategies of Husserl and Heidegger become apparent." (Okrent, cited in Dreyfus, 1992, p. 345, note 6) Even with such well-informed guiding clues, Dreyfus is dismissive and contemptuous of this view; he says that the question raised by Føllesdal is "irrelevant" (Dreyfus, 1992, p. 48) and that Husserl "never worked out a theory of action." (ibid, p. 55) It seems eminently clear that Dreyfus was completely oblivious of Husserl's *Ideas Second Book* (available in German since 1952, and translated into English in 1989), with its thorough and fine-grained analysis of action; Kristana Arp and Lester Embree (Nenon & Embree, 1996, pp. 161–171, 173–198) have presented some of Husserl's central arguments in a fashion which effectively demolishes Dreyfus' claims about Husserl's allegedly deficient idea of practice.

OVER-INTELLECTUALIZATION (INTELLECTUALISM)

(a) All forms of human comportment toward the world are construed as varieties of an intentional directedness modeled on an intellectual encounter with objective aspects of mere things; (b) the corporeal, affective, and evaluative dimensions of human being are built up out of a most basic stratum of intellectual apprehension.

As early as 1908–1914, in his Lectures on Ethics and Value-theory, Husserl had argued that apperceptions of the heart (valuations) are founded on objectifying acts, but are not reducible to those acts. In these lectures the doctrine emerges that the surround world, as one finds it in the pretheoretical, natural attitude, is always already permeated with values; that the so-called "value-free" theoretical dimension is an idealized abstraction. The thing-like status of value-free objects is the result of an abstraction from the fullness of the world, "soaked with valued objects." James Hart poses this question on behalf of Husserl's response to Heidegger's critique: "Does Husserl's position in effect hold that value perception is preceded by an absolutely prior value-free constitution of things and that values ultimately have their ontological origin exclusively in the antecedent positing of the thing-reality as the foundational layer?" (Hart, 1990, p. 214) Hart's careful unpacking of these lecture notes shows that the unequivocal answer is No. Husserl indicated as much when he used the phrase "with the same immediacy" in this passage from *Ideas Book One*: "this world is there for me not only as a world of mere things, but also with the same immediacy as *a world of objects with values, a world of goods, a practical world*." (Husserl, 1982, p. 53) Hart says that "the theme which preoccupied [Husserl] until the end of his life [was] the problem of the conative or striving nature of the primal presenting or living present." Behind all perception of objects there is the striving of the total intention, a universal or general will for which, or under which, the world is disclosed as a universal field of practice with its goals, values and preferences. (Hart, 1990, p. 215; see also Drummond, in Nenon & Embree, 1996, pp. 248–253) There is a founding pre-directedness through which consciousness orients itself, a most basic sense of will or heart that founds and "goes in advance" of the acts that constitute objects. Only if Heidegger chose to ignore these lectures could he have felt justified in leveling the charge he made against Husserl.

In the final manuscript of *Ideas Second Book*, Husserl asserted that the disclosure of structures of the surround world shows that acts of the theoretical attitude cannot always refer back to yet earlier theoretical acts; the investigation arrives at "pregiven objectivities which do not spring from theoretical acts but are constituted in intentional lived experiences imparting to them *nothing* of logico-categorial formations." He is aware of potential misunderstanding of his position about the cognitive origin of theoretical acts and says that his previous remarks need to be supplemented. "As possibilities running parallel to the theoretical attitude, there are the axiological and practical attitudes.... Valuing acts...can relate to pre-given objectivities, and their intentionality proves itself immediately thereby as constitutive for objectivities of a higher level, analogues of the categorial objectivities of the logical sphere." (Husserl, 1989, p. 9) In these preliminary passages, Husserl devotes much attention to the characteristic of all intentional acts that they can "pass over" or undergo "conversion" from one attitude to another—no attitude is privileged except in terms of its own goal. Lived experience in simple sense-intuition, performed in a theoretical attitude, grasps the mere thing; but when experience "passes over" into grasping and judging of value, "we then have more than a mere thing, we have the thing with the 'what' character of the value; we have a value thing." This grasping of the "what" character of a value-thing is not an achievement of a cognitive or intellectual intention: "each consciousness which originally constitutes a value-object as such necessarily has in itself a component belonging to the sphere of *feelings*." (Husserl, 1989, p. 11) When the phenomenologist turns from the *naturalistic* attitude, i.e., the attitude that considers human, animate bodies as parts of the natural world, toward the *personalistic* attitude, i.e., one that considers humans as persons whose words and acts are motivated by spiritual concerns, the very idea of grasping objects in the environment changes. It is not the mere "what" character of an object that may strike one's attention, but its "in-order-to": "Among the things of my environment, that one there *steers my regard* onto itself; its special form 'strikes me'. I choose the fabric *for the sake of* its beautiful color or its smoothness... In short, in my theoretical, emotional, and practical behavior... *I feel myself conditioned by the matter in question.*" (Husserl, 1989, p. 148; see also pp. 377–380) The notion of thing-like indication, its in-order-to structure, is important in Heidegger's analysis of circumspective understanding; "when an assignment to some particular "towards-this" has been thus... aroused, we catch sight of the "towards-this" itself, and along with it everything connected with the work... as that wherein concern always dwells. The context of gear [equipment] is lit up, not as something never seen before, but as a totality constantly sighted beforehand in circumspection." (Heidegger, 1962, p. 105) He describes this "whole" within which all things ready-to-hand have their assignment, their in-order-to direction, as the referential totality, like the tools in a workshop.

SPLITTING OF THE EGO

The transcendental ego is divorced from the empirical or mundane ego; human consciousness is split into two separate realms, one anonymous, lifeless, and neutral, the other, personalized, full-of-life, and interested. (cf. Heidegger, 1982a, pp. 275–276) In *Ideas First Book*, Husserl says that the results of the epoché lead toward our goal which can be characterized as "the acquisition of a new region of being never before delimited in its own peculiarity." This delimitation or exhibition shows "nothing else than what we shall designate... as 'pure mental processes', 'pure consciousness' with its pure 'correlates of consciousness' and, on the other hand, its 'pure ego'." (Husserl, 1982, pp. 63-64) Husserl here endorses the general sense of the Cartesian cogito as the most suitable platform to launch his investigation; this preeminent field is consciousness in "an extraordinarily broad sense." But he cautions that the ego that "lives in" the various mental processes must be left out of consideration, since it is spoiled or tainted with a naturalistic conception of mind as a thinking thing. But he does grant the exceptional importance of Descartes' insight into the apodictic certainty with which the ego is given in reflection on one's thoughts. Hence, Husserl argues vigorously for consciousness as absolute and indubitable, and the objects that "lie over against" it as contingent and dubitable. (Husserl, 1982, pp. 100–104) Given the logical possibility that everything outside conscious being might not exist, conscious being could survive the annihilation of the world (though it would be "modified" by such a global negation) (Husserl, 1982, pp. 110–111). This notorious thought experiment leads Husserl to a very strong conclusion: "a veritable abyss yawns between consciousness and reality. Here, an adumbrated being, not capable of ever becoming given absolutely, merely accidental and relative; there, a necessary and absolute being, essentially incapable of becoming given by virtue of adumbration and appearance." (On Husserl's "misleading manner" of presenting this, see Bernet et al., 1992, pp. 68–69; Philipse, in Smith & Smith, 1995, pp. 256–259)

Existentialist criticism, most famously by Heidegger and Sartre, has focused on one conclusion that Husserl draws from the absolute-relative distinction between non-conscious and conscious being in terms of the latter's *purity*: "consciousness considered in its 'purity' must be held to be a *self-contained complex of being*, a complex of *absolute being* into which nothing can penetrate and out of which nothing can slip." (Husserl, 1982, p. 112) It is hardly surprising that when Husserl treats the nature of this pure ego he refers to it with the Leibnizian term monad. But one should be cautious about two points in interpreting this and similar passages: first, his discussion in the section, and the pertinent distinction between inside and outside, is couched in terms of the immanent stream or flux, no "part" of which has spatial-temporal reality and no "part" of which is given in adumbrations (perceptual aspects). Second, he is concerned with the transcendental ego, the unifying and unitary pole of all conscious experience, and not with the psychical ego, constituted by the living embodied being's habitualities.

However, it goes without saying that, at the stage of *Ideas First Book*, he conceives of the pure ego as empty of content, one without any essential components, one which cannot be made the object of any study, as all other intentional "objects," including the psychical ego, can be the object of a theme. "The Ego living in mental processes is not something taken *for itself* and which can be made into an Object *proper* of an investigation. Aside from its 'modes of relation' or 'modes of comportment', the Ego is completely empty of essence components, has no explicable content, is not describable in and for itself: it is pure Ego and nothing more." (Husserl, 1982, p. 191) More than twenty years later Husserl dramatically revised his own earlier view of the pure ego; in hindsight, he now says, the so-called "Cartesian way" had a great shortcoming. "While it leads to the transcendental ego in one leap, as it were, it brings this ego into view as apparently empty of content, since there can be no preparatory explication" (Husserl, 1970, p. 155). It is this very *preparatory explication* to which he devotes so much attention in *Ideas Second Book*, about the two-layered constitution of psycho-physical human being, and in his emphasis in *The Crisis of European Sciences* on the foundational layer of human practice and willing. As late as the *Cartesian Meditations*, he still refers to the 'purity' of transcendental consciousness with its universe of "absolute freedom from prejudice." The conceptual division between an ego that has no interests and no essence from one whose interests partly comprise its essence leads to the notorious *splitting of the ego*. "If the Ego, as naturally immersed in the world . . . is called *interested in the world*, then the phenomenologically altered . . . attitude consists in a *splitting of the Ego*: in that the phenomenological Ego establishes himself as '*disinterested onlooker*', above the naively interested Ego . . . . The Ego's sole remaining interest being to see and to describe adequately what he sees, purely as seen, as what is seen and seen in such and such a manner." (Husserl, 1973, p. 35)

This statement accurately represents Husserl's attitude toward the ego as it is revealed through the reduction; but one must not forget that Husserl had already discriminated *a new sense* of transcendence and immanence in his lectures of 1906–1907 (see Bernet et al., 1992, pp. 206–208). Sartre ignores the subtle distinction between the two senses of transcendence when he criticizes Husserl's view in his short 1936 monograph on *The Transcendence of the Ego* (see Catalano, 1995; Sartre, 1973). Sartre argued that Husserl's insistence on the phenomenological reduction left only "a little residue," the subjective remainder of the process of bracketing the world's being. Such an ego, he said, is the subject that cannot be an object, and instead became a thing-in-itself, unknowable to any being, including oneself. In its place, Sartre spoke in favor of consciousness as "a pure spontaneous upsurge" that required no ego behind the scenes to hold everything together. He concluded that the transcendental ego itself must be transcended, that is, shown to be superfluous, and the empirical, everyday ego resuscitated (see especially Edie, 1994). However, this critique does not take adequate cognizance of what Husserl had to say in the Fourth Cartesian Meditation. "The ego is himself *existent for himself* in continuous evidence; thus, in himself, he is *continuously constituting himself as existing*." [In an author's later marginal note, Husserl worried that this part of the exposition might have "come too late."] "The ego grasps himself not only as a flowing life but also as I, who live this and that subjective process, who lives through this and that cogito, *as the same I*." (Husserl, 1973, p. 66) The point-of-view of the transcendental ego is only attained in reflection on one's thoughts, it cannot be completely comprised (or fulfilled) as an "object" of an intention, it is the formal-logical principle of the unification of all intentional acts, and hence it is the same ego as the mundane, ordinary ego. "To the transcendental ego there corresponds then the human Ego, concretely as the psyche taken purely in itself and as it is for itself, with the psychic polarization: I as pole of my habitualities, the properties comprised in my character." (Husserl, 1973, p. 73) In this context he stresses the notion of personal *habitus*: with every act, with every decision, with every judgement the ego acquires "an abiding property." The decision taken, the conviction acted on, and so forth contribute to the abiding ego who is thus-and-so decided; the individual becomes changed by means of these freely chosen words and deeds. (see Husserl, 1977, pp. 162–165) In later texts from the *Nachlass*, he wrote that the horizon of individual interests correlates to the general will of the person; and this general will constitutes the background decisions taken by that person which *proceed in advance* of perception, movement and choice. (Hart, 1990, p. 216)

Rudolf Bernet has advanced an argument to show that another central feature of Husserl's approach to the ego-problem, one that Heidegger explicitly repudiated, is at work in *Being and Time*—the reduction. It is through the transcendental reduction that Heidegger alleges the ego becomes split into the pure ego and the mundane ego. But in his early Marburg lectures Heidegger said that, "the phenomenological reduction is the basic component of the phenomenological method." Although he distanced himself from Husserl's transcendental *version* of the reduction, in fact, he extoled his own existential version: "For us, phenomenological reduction means leading phenomenological vision back to the apprehension of being, whatever may be the character of that apprehension, to the understanding of the being of this being (projecting upon the way it is unconcealed)." Bernet argues that not only is the reduction at work in *Being and Time*, but that as Heidegger construes his own version, it is not that different from Husserl's more mature version, one worked out with his assistant Eugen Fink. In the Sixth Cartesian Meditation, the task of the reduction is "to bring to light the 'transcendental' correlation between constituting consciousness and the constituted world. And it is true that making this correlation manifest also means showing that it was already at work in the anonymity of 'natural' life without natural life realizing it. But this in no way gives Heidegger the right to say, as he does repeatedly, that Husserl is not interested in the issue of the being of constituting consciousness. On the contrary, by revealing the correlation between constituting consciousness and the constituted world, the phenomenological reduction makes manifest precisely the (pre-)being of this consciousness and the being of this world as well as the difference between them." (Bernet, in Kisiel & van Buren, 1994, p. 255) Merleau-Ponty observed that "far from being . . . a procedure of idealistic philosophy, phenomenological reduction belongs to existential philosophy; Heidegger's being-in-the-world appears only against the background of the phenomenological reduction." (Merleau-Ponty, 1962, p. xiv)

Bernet's exegesis is not designed to show that Heidegger's use of the reduction is *the same as* Husserl's use, nor that their conception has the same ground and orientation. For Heidegger, Dasein's natural life assumes the form of circumspective concern with available items in its familiar environment. In the ordinary course of events Dasein does not pay attention to the way in which equipment signifies relations within the world or to its own concern existence, nor to the way in which this existence inserts itself into the world. When natural life malfunctions and things go wrong the *first* reduction reveals the correlation between Dasein's inauthentic or improper existence and the world to which it and the items it uses refer. In line with the Sixth Cartesian Meditation (Fink & Husserl, 1995), this first existential reduction "makes manifest the equipment's intrawordly being in its relation to the being-in the-world of concern in Dasein, reveals the hidden being of the world and Dasein, their difference and their bond." The revised Husserlian version of the transcendental reduction takes another step in elucidating the being of the subject by asserting that the subject's life is not limited to its constituting activity. By bringing out the hidden work of the world's constitution, the next step "effects a split within this subject"; it is both the constituting subject and an observer of the activity of this constitution, an observer that does not pre-exist in its natural life. "While one may speak of constituting consciousness operating unknowingly within natural life, there is no phenomenological spectator that does not know itself, that is, does not know what it is doing." (This same point regarding the Sixth Cartesian Meditation is made by Merleau-Ponty, 1962, p. xx) Bernet also argues that the revised Husserlian reduction provides an insight into another distinctive existential theme—the *leap into a higher realm*—through the difference uncovered between worldly beings and the being of constitutive consciousness. The ontological difference between the two "requires a 'leap' (*sprung*)," Bernet says; "But this leap out of natural life is ventured, dared by the subject who *wants* to know more, who *wants* to have a clear mind, who *wants* to give a verdict in the name of scientific evidence. There is nothing of this sort in Heidegger." (Bernet et al., p. 258)

Consciousness and World Separated by an Abyss

The whole world is sundered from consciousness, separated by an abyss in being, since the being of the conscious is radically different from the being of the non-conscious. Consciousness is separated from the world of the factually living human by a gulf. (Heidegger, 1982a, pp. 156–158, 1982b, p. 98) The purchase of this objection is only as good as the probity of the previous objection about the split in the ego, and falls before the same Husserlian definition of transcendence. In his 1925 Lectures on Phenomenological Psychology, Husserl stresses the point that nature and mind (as he calls it here) are *not* sundered by an abyss, not radically opposed to each other, but intertwined or interwoven together. "The natural and the mental do not confront us clearly and separately so that mere pointing would suffice: *here* is nature, and *here*, as something completely different, is mind. Rather, what seems at first obviously separated, upon closer examination turns out to be obscurely intertwined, permeating each other in a manner very difficult to understand." The obviousness of their separation is the unfortunate consequence of a natural scientific attitude toward the nature of mind. "As scientific themes, nature and mind do not exist beforehand; rather, they are formed only within a theoretical interest... upon the underlying stratum of a natural, pre-scientific experience. Here they appear in an originally intuitable intermingling and togetherness." (Husserl, 1977, pp. 39—40) In the Fourth Cartesian Meditation, he is positively contemptuous of the kind of spuriously scientific philosophy that conceives the world in its objective status as literally "over-against" consciousness. "The attempt to conceive the universe of true being as something lying outside the universe of possible consciousness, possible knowledge, possible evidence, the two being related to one another merely externally by a rigid law, is nonsensical. They belong together essentially and... they are also concretely one." (Husserl, 1973, p. 84) One of Husserl's main lines of defense against the allegation that mind and world are separated by an abyss or a gulf is that his idea of the being of the intentional agent has been misunderstood.

Neglect of the Being of the Intentional

The issues of the being of the intentional act and the being of an intentional agent are neglected. (Heidegger, 1982a, pp. 65, 115, 1982b, pp. 114–115) Daniel Dahlstrom has opened a line of inquiry about whether Heidegger fairly presented Husserl's thought on intentionality and inner time consciousness, and whether he actually overcomes and corrects these alleged failures. Heidegger portrayed Husserl as the Moses of traditional philosophy, "at once its liberator and its victim, pointing the way out of the desert through a clarification of intentionality, but unable to enter the promised land of existential analysis." (Dahlstrom, in Kisiel & van Buren, 1994, p. 231) According to Heidegger in these early lecture notes, Husserl's phenomenology leaves unanswered the question of the ontological status of intentionality, or more properly, intentional being, and the meaning of being as such, both themes central to Husserl's exposition of the primary reality of consciousness. According to his protégé's critique, Husserl accepts and then employs a traditional prejudice that separates at the deepest level the being of consciousness and the being of particular existents. "Husserl's specific characterization of being in a primary sense, as what is absolutely given in pure consciousness, is based on an attempt to elaborate, not what 'to be' means, but rather what is necessary for consciousness to constitute an 'absolute science'. For the phenomenologist above all, the failure to raise the question of what 'to be' means is of a piece with a failure to unpack what 'to be' means in the case of a particular sort of being (*Seiendes*), namely, consciousness, understood as 'intentionality.' This twin failure is, moreover, the direct result of an infidelity to phenomenology's most basic principle." (ibid, p. 237)

With regard to phenomenology's cardinal directive, "to the things themselves," Heidegger alleges that Husserl determines the meaning of 'thing' according to a traditional preconception; in doing so Husserl fails to bracket the naturalistic concept of 'being' and reduces consciousness to a natural being in the world, instead of an existential being-in-the-world. Dahlstrom asks, "How fair is Heidegger's criticism? On the one hand, despite his enthusiasm for Husserl's intentional analysis of knowing, Heidegger's criticism effectively discounts the essential role Husserl accords the mere, or even empty, intending of things in the constitution of primary significance of 'truth' and 'being'. Yet this fundamental feature of Husserl's analysis belies the reproach that he crudely equates being with sheer presence. Heidegger's critical exposition of Husserl's phenomenology is, moreover, highly selective, ignoring several other nuances and details of its analyses and development. On the other hand, ... there are certainly grounds for Heidegger's contention that the basic structure of objectifying acts or, more specifically, an ontology of presence dominates the horizon against which accounts are given of truth, being, and intentionality." (ibid, p. 239)

In attempting to overcome Husserl's supposed view of "objects" as lying over against consciousness, Heidegger deploys another formal schema, the temporal horizon of human being, without acknowledging Husserl's long-running, meticulous investigations of inner time. Dahlstrom responds on Husserl's behalf against Heidegger's spurious claims: "Husserl plainly addresses the ultimate horizon for objectifying acts that is accordingly itself neither an object nor an act. There is still a kind of consciousness involved, but it is not the objectifying consciousness of intentionality that Heidegger criticized as prejudicially scientific; if anything, internal time consciousness is prescientific, even 'preobjective,' starkly contrasting with the sketch Heidegger gives in his lectures of Husserl's account of intentionality. Nor is Heidegger unaware of this significance. In the preface to the published part of these manuscripts he writes that what is particularly telling about them is 'the growing, fundamental clarification of *intentionality* in general', a remark that must have surprised the students who listened to him during the summer semester of 1925." (ibid, pp. 239–240) Various reasons have been advanced for why the younger philosopher remained silent about the elder thinker's outstripping these criticisms; but irrespective of the motive or rationale, it is clearly the case that Heidegger both misrepresents Husserl's position and obviates the grounds for his disagreement.

SUMMARY

In summary, for Husserl human being has an underlying basis in its lived body and its habitus; it is this basic stratum that individuates each person in his or her individuality, and, as such, the human person does *not* have an essence in advance, but instead is underway, a being open to its future. "The pregivenness of the world signifies the persistency of a universal world-conviction, a world-possession, which at the same time is a presumption of being and is always givenness of being, indeed as givenness of a being which has *its true being only ahead of itself*." (Husserl, 1989, p. 363) In *The Crisis of European Sciences*, Husserl clearly links his concept of the individual not having an essence in advance, an embodied being that is underway and always becoming, to the core concept of an internal, self-realized freedom—and, in doing so, clearly captures the Existentialist concept of freedom. (see esp. Edie, 1984) "This life, as personal life, is a constant becoming through a constant intentionality of development. What becomes, in this life, is the person himself. His being is forever becoming. (Husserl, 1970, p. 338) And when Husserl does reflect on the possible grounds which could motivate the philosopher to perform the phenomenological reduction he refers to "the irrational fact of the world's rationality" (Bernet et al., 1992, p. 10), and the "clarification of the world leads back to the task of systematically setting out the final irrationality of the world-constituting consciousness." (Husserl, in Nenon & Embree, 1996, p. 12) This startling remark gestures toward the radical contingency at the basis of human existence, an emphatically Existentialist notion. Husserl's immediate response to the challenge posed by the lack of inherent reason, namely, the resolve to carry through with the reduction and all that this entails, indicates a philosophical calling followed under the title of "my fullest freedom."

REFERENCES

Bernet, R., Kern, I., & Marbach, E. (1992). *An introduction to Husserl's phenomenology*. Evanston, IL: Northwestern University Press.

Catalano, J. S. (1995). Reinventing the transcendental ego. *Man and World, 28*(1), 101–111.

Dreyfus, H. (1992). *Being in the World*. Cambridge, MA: MIT Press.

Edie, J. M. (1984). The roots of the existentialist theory of freedom in ideas II." *Husserl Studies, 1*, 243–261.

Edie, J. M. (1994). The question of the transcendental ego: Sartre's critique of Husserl. In T. Stapleton (Ed.), *The Question of Hermeneutics*. Dordrecht: Kluwer Academic.

Fink, E., & Husserl, E. (1995). *Sixth cartesian meditation*. Bloomington: Indiana University Press.

Føllesdal, D. (1979). Husserl and Heidegger on the role of actions. In E. Saarinen (Ed.), *Essays in honor of Jaakko Hintikka* (pp. 365–378). Dordrecht: D. Reidel.

Hart, J. (1990). Axiology as the form of purity of heart. *Philosophy Today*. vol. 34, 206–221.

Heidegger, M. (1962). *Being and time* (E. Robinson & J. Macquarrie, Trans.). NY: Harper and Row.

Heidegger, M. (1982a). *The basic problems of phenomenology* (A. Hofstadter, Trans.). Bloomington: Indiana University Press.

Heidegger, M. (1982b). *History of the concept of time* (T. Kisiel, Trans.). Bloomington: Indiana University Press.

Heidegger, M. (1998). *Pathmarks* (W. McNeill, Ed. & Trans.). Cambridge: Cambridge University Press.

Husserl, E. (1970). *The crisis of european sciences* (D. Carr, Trans.). Evanston, IL: Northwestern University Press.

Husserl, E. (1973). *Cartesian meditations* (D. Cairns, Trans.). The Hague: Nijhoff.

Husserl, E. (1977). *Phenomenological psychology* (J. Scanlon, Trans.). The Hague: Nijhoff.

Husserl, E. (1982). *Ideas first book* (F. Kersten, Trans.) Dordrecht: Kluwer Academic.

Husserl, E. (1989). *Ideas second book* (R. Rojcewicz & A. Schuwer, Trans.). Dordrecht: Kluwer Academic.

Husserl, E. (1997). *Psychological and transcendental phenomenology and the confrontation with heidegger* (T. Sheehan & R. Palmer, Trans.). Dordrecht: Kluwer Academic.

Kisiel, T., & van Buren, J. (Eds.). (1994). *Reading Heidegger from the start: essays in his earliest thought*. Albany, NY: State University of New York Press.

Levinas, E. (1973). *The theory of intuition in Husserl's phenomenology* (A. Orianne, Trans.). Evanston, IL: Northwestern University Press.

MacDonald, P. S. (2000). *Descartes and Husserl*. Albany, NY: State University of New York Press.

Merleau-Ponty, M. (1962). *Phenomenology of perception* (C. Smith, Trans.). London: Routledge and Kegan Paul.

Merleau-Ponty, M. (1996). In H. Silverman & J. Barry, Jr., (Eds.), *Texts and dialogues* (2nd ed.). NJ: Humanities Press.

Moran, D. (2000). Heidegger's critique of Husserl's and Brentano's accounts of intentionality. *Inquiry, 31*(1), 39–66.

Nenon, T., & Embree, L. (Eds.) (1996). *Issues in Husserl's ideas II*. Dordrecht: Kluwer Academic.

Sartre, J.-P. (1973). *The transcendence of the ego: An existentialist theory of consciousness* (F. Williams & R. Kilpatrick, Trans.). NY: Farrar, Strauss and Giroux.

Smith, B., & Smith, D. W. (Eds.) (1995). *Cambridge companion to Husserl*. Cambridge: Cambridge Univ. Press.

van Buren, J. (1994). *The young Heidegger: Rumor of the hidden king*. Bloomington: Indiana University Press.

CHAPTER 7

THE INFLUENCE OF HEIDEGGER ON SARTRE'S EXISTENTIAL PSYCHOANALYSIS

MILES GROTH

In this chapter, I will address two questions: What was Sartre's contribution to psychology and to what extent was Sartre's psychology influenced by Heidegger's thought? I will concentrate on Sartre's existential psychoanalysis as outlined in *Being and Nothingness* (Sartre, 1956). After some preliminaries, I give an account of Sartre's existential psychoanalysis and then, in the third section of the chapter, review the intellectual encounter between Heidegger and Sartre, especially as it bears on the early Sartre's alternative to the several incarnations of empirical psychoanalysis, which began with Freud.

PRELIMINARIES

Much like what will go on between an analysand and her psychoanalysis, the relation between Jean-Paul Sartre's thought and psychoanalytic theory, once set in motion, was to be ambivalent and lifelong. In a late interview, Sartre (1974a, p. 36) reported: " [I] had a deep repugnance for psychoanalysis in my youth." He goes on to explain that, when he was a young French intellectual, psychoanalysis was considered to be an example of "'soft' thought" (p. 38), that is to say, lacking in Cartesian rigor and dialectical subtlety. Nevertheless, in his late thirties, Sartre published a book, *Being and Nothingness*, that contained a section entitled "Existential

Psychoanalysis." Published 10 years later in an English translation under that title, the text (along with another section from *Being and Nothingness*, "Bad Faith") was for some English-speaking readers regarding their introduction to Sartre's most influential philosophical work, and, apart from *Existentialism and Humanism,* very likely their first exposure to Sartre's philosophy, assuming they did not have in hand a copy of the "special abridged version" of *Being and Nothingness.*[1]

Sartre's existential psychoanalysis is offered as an alternative to empirical psychoanalysis. As applied by Sartre to historical figures such as Charles Baudelaire (Sartre, 1967) and Gustave Flaubert (Sartre, 1971) existential psychoanalysis has much in common with Erik Erikson's psychohistories (Erikson, 1958, 1969).[2] While Sartre presents examples of the existential analysis of an individual's moods and interests in "Existential Psychoanalysis," he does not, however, offer technical advice on how to work with individuals in the psychoanalytic setting and never practiced as a psychotherapist.

Sartre's interest in academic psychology was straightforward. As a young man, writing under the influence of Husserl's phenomenological psychology, he made four important contributions to the field: a study of consciousness, *The Transcendence of the Ego* (Sartre, 1957); two books on the imagination, *The Imagination* (Sartre, 1936) and *The Imaginary. Phenomenological Psychology of the Imagination* (Sartre, 1948a); and a monograph entitled *Outline of a Theory of the Emotions* (Sartre, 1948b).[3] These studies are clearly in the background of Sartre's existential psychoanalysis, and his critique of empirical psychoanalysis is already evident in them.[4] By 1960, in *Search for a Method* (Sartre, 1963b) and the *Critique of Dialectical Reason* (Sartre, 1976) psychoanalysis was given only occasional mention, and got mixed reviews by Sartre.[5] With his concerns now directed more to social life, thanks to his reading of Marx, Sartre may seem to have lost interest in the psychoanalytic study of the individual.[6] Yet, his last great (uncompleted) work on Flaubert is another exercise in existential psychoanalysis.

But if Sartre's interest in Freudian psychoanalysis was cooling off as he matured intellectually, what are we to make of his participation, beginning in 1958, with the movie director John Huston in the preparation of a screenplay entitled *Freud* (Sartre, 1985)? Some may regard it as Sartre's existential psychoanalysis of Freud, comparable to his other literary existential psychoanalyses.[7] In any event, in the work of the early Sartre, with whom I am concerned here, interest in empirical psychoanalysis was still so strong, albeit highly critical, that he developed an alternative to it. What is existential analysis?

SARTRE'S EXISTENTIAL PSYCHOANALYSIS

The doctrine of Sartre's existential analysis begins with his observation that human reality is a unity, not an ensemble of faculties. "The *principle* of this psychoanalysis is that man is a totality and not a collection" (*BN*, p. 568) of functions.[8] As such, a person cannot be pieced together from the elements discovered by an analysis of his psychological functions (perception, cognition, and will), his memories reassembled as a psychobiography, or even the details of his stages of development. In fact, according to Sartre, when understood with discernment, each element *is* the unity.

Sartre's existential psychoanalysis is grounded in his ontology of human reality, which he formulates in the following way: "Fundamentally man is the desire to be," albeit "the existence of this desire is not to be established by an empirical induction" (*BN*, p. 565). Instead, the desire to be can be discovered only by tracing back what a person has done to his original choice of what to be (his project), that is, to "that original relation to being which constitutes the person who is being studied" (*BN*, p. 595). Regrettably, the desire to be something—in other words, "the desire of being-in-itself" (*BN*, pp. 565–566)—always falls short of its goal. Instead, the person is limited by his status as a being-for-itself. This means that he continually overreaches himself and is never in a position to see or know himself as the unity he is, since he is forever changing and always just eludes even a glimpse of himself as a whole. This is an inevitable state of affairs, given that being-for-itself is the ontological status of human reality (existence), which Sartre defines as freedom.

For human reality, as the famous assertion runs, "existence precedes essence" (*BN*, p. 568), and "far from being capable of being described," for example, "as *libido* [by Freud or Jung] or will to power [by Adler], [human reality] is a *choice of being*, either directly or through appropriation of the world" (*BN*, p. 602). Crediting "Freud and his disciples" with a "first outline of this method" (*BN*, p. 599) of revealing the meaning of human reality, Sartre nevertheless suggests that empirical psychoanalysis must give way to a psychoanalysis that "recognizes nothing *before* the original upsurge of human freedom" (*BN*, p. 569).

Existential analysis also denies the notion of the unconscious any meaning and "makes the psychic act coextensive with consciousness" (*BN*, p. 570). When existential psychoanalysis is effective, the analyst or analysand merely comes "to know what he already understands" (*BN*, p. 571). He does not discover previously inaccessible unconscious contents, which were by definition beyond his understanding.

Just as he dismisses the meaningfulness of unconscious mental life, Sartre also rules out the unmediated influence of the environment on an individual's comportment. "The environment," he writes, "can act on the subject only to the exact extent that he comprehends it; that is, transforms it into a situation" (*BN*, p . 572). As for what the analysand reports that can be interpreted symbolically, Sartre precludes recourse to universal or collective symbols. He explains: "If each being is a totality, it is not conceivable that there can exist elementary symbolic relationships ... we must always be ready to consider that symbols change in meaning" (*BN*, p. 573).

Finally, for Sartre, there is no fixed technique or method of existential psychoanalysis. "The method which has served for one subject will not necessarily be suitable to use for another subject or for the same subject at a later period"

(*BN*, p. 573). Unfortunately, Sartre does not illustrate or detail how such a method might work in the traditional therapeutic setting.

In general terms, existential psychoanalysis "is a method destined to bring to light, in a strictly objective form, the subjective choice by which each living person has made himself a person; that is, makes known to himself what he is" (*BN*, p. 764). Its goal is to discover in the series of involvements comprising one's existence "the original way in which each man has chosen his being" (*BN*, p. 599). Nothing further is required of the analysand.

From a philosophical perspective, existential psychoanalysis picks up where ontology leaves off; moreover, existential psychoanalysis depends upon ontology. "What ontology can teach psychoanalysis is first of all the *true* origin of the meaning of things and their *true* relation to human reality" (*BN*, p. 603). In a discussion of desire, Sartre explains in more detail the relation between ontology and existential psychoanalysis:

> Existential psychoanalysis can be assured of its principles only if ontology has given a preliminary definition of the relation of these two beings—the concrete and contingent in-itself or object of the subject's desire [a concrete existent in the midst of the world], and the in-itself-for-itself [the desire to be a certain something or someone] or [alas, he admits, impossible[9]] ideal of the desire—and if it has made explicit the relation which unites appropriation as a type of relation to the in-itself, to being, as a type of relation to the in-itself-for-itself (*BN*, pp. 585–586).

In other words, before existential psychoanalysis can begin, ontology must have articulated the relation of a subject's desire to be a certain something, both to the object of his desire and to being.

A distinctive feature of Sartre's view of human reality emerges in his discussion of desire; namely, the significance of *possession* for human reality. For Sartre, desire invariably tends to possession of its object, and "in a great number of cases, to possess an object is to be able to use it" (*BN*, p. 586). This holds for our relation to things as well as to other human beings. Moreover, the "internal, ontological bonding between the possessed and the possessor" (*BN*, p. 588) extends to wanting to *be* the object of desire, so that, in the end, "I *am* what I have" (*BN*, p. 591) and have come to possess. Sartre describes the characteristics of possession:

> Possession is a magical relation: I *am* these objects which I possess, but outside, so to speak, facing myself; I create them as independent of me; what I possess is mine outside of me, outside all subjectivity, as an in-itself which escapes me at each instant and whose creation at each instant I perpetuate (*BN*, p. 591).

And so it happens that, whether in relation to my computer keyboard or my lover, "I am always somewhere outside of myself" (*BN*, p. 51).

Finally, for Sartre, desire for anything in particular is at the same time desire for the world in its entirety, so that "to possess is to wish to possess the world across a particular object" (*BN*, p. 597).

Commenting on the usual explanation of how perception operates in desire, Sartre reminds his reader that "[t]he yellow of the lemon ... is not a subjective mode of apprehending the lemon; it *is* the lemon" (*BN*, p. 603). Similarly, the beauty of someone I desire is not merely a "subjective mode" of my experience of the person; it *is* the person. In a comment from which developmental psychology might learn a great deal, Sartre suggests that, because "we come to life in a universe where feelings and acts are all charged with something material . . . and in which material substances have originally a psychic meaning" (*BN*, p. 605), the objective quality of a thing "is a new *nature* which is neither material (and physical) nor psychic, but which transcends the opposition of the psychic and the physical, by revealing itself to us as the ontological expression of the entire world" (*BN*, p. 606). Thus, for human beings, all things are human things at first—the rattle or blanket, just as much as the mother's breast or nose—but gradually some things acquire non-human status, while the others go on being human.[10]

In practice, the work of existential psychoanalysis is "to reveal in each particular case the meaning of the appropriate synthesis [of having and being]" for which ontology provides "the general, abstract meaning" (*BN*, p. 595). A kind of psychoanalysis that "must bring out the *ontological* meaning of qualities" (*BN*, p. 599), existential psychoanalysis therefore relies on "the things themselves, not upon men" (*BN*, p. 601), including the words they use.

In *Being and Nothingness*, the dialogue with empirical psychoanalysis is foreshadowed in an earlier section of the work called "Bad Faith."[11] *Mauvaise foi* (literally, falsehood) is "a lie to oneself" (*BN*, p. 49). It is a phenomenon of consciousness that must be understood against the background of Sartre's view that "the being of consciousness is the consciousness of being" (*BN*, p. 49). Bad faith is unique, in that it is a "metastable" psychic structure; that is, it is "subject to sudden changes or transitions" (*BN*, p. 50, n. 2). *Mauvais foi* is ubiquitous. It is "a type of being in the world, like waking or dreaming, which by itself tends to perpetuate itself" (*BN*, p. 68).

Having recourse to the notion of unconscious mental life, empirical psychoanalysis "substitutes for the notion of bad faith, the idea of a lie without a liar" (*BN*, p. 51). For Sartre, a close analysis of Freud's psychodynamic model reveals that the censor operating in the psychic apparatus is in bad faith (*BN*, p. 53). He believes that introducing the censor into the psychic apparatus allowed Freud to make bad faith part of our psychic structure, rather than a pathological feature of psychological functioning.

Sartre illustrates bad faith in action and refines his conceptualization of the phenomenon by contrasting it with its "antithesis," sincerity. In a discussion of bad faith and sincerity, he stresses that the laws of logic do not hold for human reality as they do for things "in" nature. Specifically, the presumed universal law of identity (A = A) fails to apply to us, since we never are what we are and always are what we are not. This applies to interpersonal relations as well, since consciousness of the other is also never what it is (*BN*, p. 62).

What is the point of posturing? The result of bad faith is "[t]o cause me to be what I am, in the mode of 'not being what one is,' or not to be what I am in the mode of 'being what one is'" (*BN*, p. 66). One of the primary goals of existential psychoanalysis is to uncover the source of bad faith, in order to reveal the truth of one's original project.

With this brief review of Sartre's exposition of existential psychoanalysis in mind, what are we to make of his alternative to empirical psychoanalysis?[12] Perhaps the most significant difference is its focus on the present. This does not preclude the goal of existential psychoanalysis to discover the fundamental project governing an individual's life, which is accomplished by means of an examination of current involvements that look back to the decisive choice that set in motion the life they comprise.

By contrast, the archeological method of empirical psychoanalysis painstakingly reconstructs the analysand's past. Some will call its result a biography; others will construe it as a narrative or story. The French word *histoire* captures both senses of reconstruction in empirical psychoanalysis. The point is, a reconstruction need not relate to the truth. It is of little importance (to the analyst or analysand) whether the analysand's personal reality as reconstructed in analysis corresponds to the historical record that is construed and maintained by the official observers of consensual reality in which the analysand grew up—first and not least the analysand's parenting figures, but also other observers of the analysand's life and times.

A second difference between the two forms of psychoanalysis follows from the first and refers back to the ontological grounding of existential psychoanalysis. That is its emphasis on freedom, in contrast to the fatalistic, solid determinism of classical empirical psychoanalysis. For the early Sartre, one can choose a different path at any time. Obviously, this does not mean one can choose *any* path. Thus an amputee cannot in good faith choose to train as sprinter, nor can a male choose to bear a child.[13] But among the feasible paths, one may choose any of them.

From a therapeutic point of view, according to orthodox empirical psychoanalysis, once in analysis, always in analysis, so that, from a psychodynamic perspective, any analysis is interminable. Sooner or later, meeting at regular sessions ends and gives way to a style of self-reflection that haunts the analysand for the rest of her life. By contrast, the end of Sartrean existential psychoanalysis has been reached when the analysand has tracked down and apprehended the event that set her on the course she *is following*, which (to use a grammatical metaphor) may be thought of as her life as lived in the middle voice ("am being"), albeit understood in terms of the future perfect tense ("will have done").

Two considerations of interest follow from the aims of existential analysis.

1. I cannot escape the fact that I never *am* anything. To frame the assertion in the language of developmental psychology, identity is a myth. The much vaunted inner sense of being more or less the same person over time is a fiction. For example, since I always "am being," I will never attain the sense of ego identity that is said to be the goal of adolescence.[14] Instead, in quest of being (what I project for myself), I am forever about becoming who I am. My non-status as a being-for-itself may be unsettling, but it is the privilege and burden of embodying freedom.

The tightrope tension of living my life in the middle voice also deprives me of the certainties that apply to a being with a fixed nature (for example, my cat), so that, in the absence of "having" any such nature, the construction of unambiguous rules or principles to guide or adjudicate the conflicts that arise in my life will be a futile endeavor. My status also exempts me from being predictable, which is one of empirical psychology's coveted goals.[15] At best, I may be kept on a more or less even keel by a morality of ambiguity, such as the one proposed by Simone de Beauvoir around the time Sartre developed his phenomenological ontology of human reality.[16]

Like time, I seem to flow on, and to observe what I would like to think I finally have become is impossible, since that could take place only when I am no longer in a position to see or know anything at all, that is, after my death. It is true that I will then be firmly in the summarizing gaze of others, now chiefly mourners (or those relieved by my passing), but, as Sartre has made clear, that has always been the case.

If I attempt to fix my present for an examination of the stretch from a given starting point to the present moment, I find that the present moment has already passed. This may trouble me (and it should, according to Sartre), but it need not bring me to a psychoanalyst's office. That happens when I am at sea about where I am heading, which is based on the fact that I have lost sight of my fundamental project.

2. A second issue that existential psychoanalysis raises is related to the first and takes the form of a question I may ask myself. Although, Sartre does not discuss the question, it has fascinated me since first encountering his work and, I think, a few words of discussion of the question may shed some light on his existential psychoanalysis. Moreover, it bears on my later discussion of Sartre's understanding of Heidegger's notion of *Existenz*.

   Assume that I have studied piano for many years. I began, when I was a boy, 8-years-old. I took a half-hour lesson every Saturday morning for ten years, then continued to practice technique and learned the piano repertoire on my own. I have spent an hour or so at the piano nearly every day since my first lesson. Now, more than 40 years have passed, and I am sitting at a computer keyboard writing an essay on Sartre's existential psychoanalysis. The question arises: Am I pianist? Strictly speaking, I am not. Perhaps, I will take a break in an hour or so and, providing it is not too late in the day (so as not to bother the neighbors!), I will open *The Well-Tempered Clavier* of Bach and play the C major prelude of Book One. As I play, I am a pianist, but only then. And while playing, am I writer? Clearly, I am not. The impossibility of *being* anything at all suggested by the first issue is in this way complicated by a further ambiguity. It refers to a different ontological issue from the one that concerns Sartre, but it is of interest, I think, for the practice of Sartrean existential psychoanalysis.

To address the second issue from the perspective of the first, I might say that, because I never *am* anything, I could never have become a pianist. I have been asking a meaningless question. Thus, to say that I am a pianist, albeit only while playing the instrument, would be unacceptable to Sartre, since to be defined as anything at all is a betrayal of one's freedom. However, the question bears on the meaning of my *Existenz*, Heidegger's term for the way of life an individual is enacting at any given moment—political *Existenz* (as I protest government policy), philosophical *Existenz* (as I think about Sartre's ontology),artistic *Existzenz* (as I play the piano).

HEIDEGGER'S INFLUENCE ON SARTRE'S EXISTENTIAL PSYCHOANALYSIS

From early on in his career, Sartre's relation to Heidegger's thought was just as important as his intellectual affair with psychoanalysis.[17] The story of Sartre's reading of Heidegger includes the attribution to Heidegger of the title existentialist, which, of course, Heidegger denied. The text that reflects his disavowal is his "Letter on 'Humanism'," which was written in response to a letter, dated November 11, 1946, from a young French philosopher, Jean Beaufret.[18] As we will see, given its title and by Heidegger's own acknowledgment in the text, the "Letter" was written with Sartre's lecture "L'Existentialisme est un Humanisme" very much in mind.

I begin with Sartre's lecture. He gave the talk at a meeting of the Club Maintenant, at Centraux Hall in Paris.[19] It is "a defense of existentialism" (*EH*, p. 23) against its opponents, in particular French Communists and Catholics. Existentialism, he writes, is "a doctrine . . . which affirms that every truth and every action imply both an environment and a human subjectivity" (*EH*, p. 24). He notes that "there are two kinds of existentialists: . . . the Christians [Jaspers and Marcel] . . . [and] the existential atheists, amongst whom we must place Heidegger as well as the French existentialists and myself" (*EN*, p. 26). Referring to Heidegger's characterization of man as "the human reality" (presumably translating Heidegger's *terminus technicus Dasein*), Sartre formulates the defining belief said to be shared by all existentialists; namely, that "*existence* comes before *essence,*" which he qualifies with the explanatory gloss, "or, if you will, we must begin from the subjective" (*EH*, p. 26). Then comes the frequently quoted passage:

> . . . man first of all exists, encounters himself, surges up in the world—and defines himself afterwards. If man as the existentialist sees him is not definable, it is because to begin with he is nothing. He will not be anything until later, and then he will be what he makes of himself. Thus, there is no human nature, because there is no God to have a conception of it. Man simply is. . . . Man is nothing else but that which he makes of himself (*EH*, p. 28).

The human reality is that we are poised in the present, but future-bound rather than determined by the past. Leaving us bereft of any appeal to excuses generated in consequence of what has happened to us or what we have done in the past, "the first effect of existentialism is that it puts every man in possession of himself as he is, and places the entire responsibility for his existence squarely upon his own shoulders" (*EH*, p. 29), and not, for example, on those of his parents (as Freud's empirical psychoanalytic theory claims) or on the effects of the environment (as the behaviorists would have it).

The ethical dimension or "deeper meaning" of existentialism is immediately evident in Sartre's subsequent observation that "in choosing for himself he chooses for all men" (*EH*, p. 29) what it means to be human. The meaning of what it is to be human is being invented by each of us every time we make a choice. Using the image of a sculptor, Sartre writes: "In fashioning myself, I fashion man" (*EH*, p. 30). Likening the human being to a painter: "[I]n life, a man . . . draws his own portrait and there is nothing but that portrait" (*EH*, p. 42).

The value-saturated state of human reality is the source of our feelings of "anguish, abandonment and despair" (*EH*, p. 30).[20] As such a state of affairs, human reality "*is* freedom" (*EH*, p. 34), and as we know from the ontological grounding of Sartre's existential psychoanalysis, freedom is not merely one of our characteristics, but our defining status. As radical freedom, "[w]e are left alone, without excuse." Thus, "from the moment that he is thrown into the world [a man] is responsible for everything he does" (*EH*, p. 34). His responsibility is not merely personal, but implicates everyone else (strangers as much as those who know what we do), and we are in the disquieting position of being without any certitudes on which to base our judgments. Sartre concludes that the "despair" of being an existentialist lies in the fact that "I remain in the realm of possibilities" (*EH*, p. 39), not of determinate actualities, let alone certainties.[21]

Sartre cautions that these givens of *la condition humaine* should not suggest an attitude of passivity or "quietism." Quite to the contrary: "The doctrine I am presenting before you," he says, "is precisely the opposite of this, since it declares that there is no reality except in actions" (*EH*, p. 41). We are, in practice and effect, nothing but a series of "purposes" (*EH*, p. 41) or "undertakings" (*EH*, p. 42). As such, existentialism must be construed as a philosophy of radical and uncompromising optimism.

Sartre sums up the philosophical rationale of existentialism in three propositions: (1) ". . . we [existentialists] seek to base our teaching upon the truth, and not upon a collection of fine theories"; (2) "this theory [sic!] alone is compatible with the dignity of man, it is the only one which does not make man into an object"; and (3) "although, it is impossible to find in each and every man a universal essence that can be called human nature, there is nevertheless a human universality of *condition*" (*EH*, pp. 44–46)—*la condition humaine*.[22]

Moving on in the lecture to the question whether existentialism is a form of humanism, Sartre distinguishes between two senses of the term "humanism," which may denote (1) "a theory which upholds man as the end-in-itself and as the supreme value" (*EH*, p. 54).[23] However, (2) its "fundamental meaning," says Sartre, "is this: Man is all the time outside of himself: it is in projecting and losing himself beyond himself that he makes man to exist" (*EH*, p. 55). Understood in this sense, humanism, like existentialism, is a doctrine of transcendence.

Human reality is ever "self-surpassing, and can grasp objects only in relation to his self-surpassing" (*EH*, p. 55). Existential humanism can therefore claim that subjectivity (understood in the sense of "man who is not shut up in himself but forever present in a human universe" [*EH*, p. 55]) and transcendence (self-surpassing) are the same. Finally, as we recall, existentialism, in particular as applied in existential psychoanalysis, begins with subjectivity. So does humanism, as Sartre understands it, and, in this sense, the two positions may be construed as equivalent.

The references to Heidegger in Sartre's lecture are obvious, though unacknowledged for the most part, and Heidegger is mentioned by name only once, as an example (with Sartre) of an atheistic existentialist. The important question now before us is whether Sartre's Heidegger reflects what Heidegger says, to say nothing of what he thought.[24]

Those who heard or read Sartre's words concluded that he and Heidegger were of one mind. Heidegger soon corrected Sartre's misunderstanding of his thought in his "Brief über den 'Humanismus'" (Heidegger, 1976, pp. 313–364).[25] Sartre is mentioned by name several times in the text; for example, where Heidegger explains that

> Sartre articulates the fundamental statement of existentialism in this way: life [*Existenz*] precedes what is by nature [*Essenz*]. In doing so, he takes (the terms) *existentia* and *essentia* in the sense they have for metaphysics, which since Plato has said that *essentia* [as possibility] precedes *existentia* [as actuality]. Sartre reverses this proposition (*HB*, 328).[26]

Explicitly disassociating himself from Sartre, Heidegger continues:

> Sartre's main point about the priority of *existentia* over *essentia* justifies the word 'existentialism' as a suitable name for this philosophy. But the main point of "Existentialism" has not the least bit in common with the sentence from *Being and Time* (cited earlier: "The 'essence' of existence [*Dasein*] lies in its way of life [*Existenz*]" and used by Sartre to characterize Existentialism). To say nothing of the fact that, in *Being and Time*, no statement about the relationship between *essentia* and *existentia* can even be expressed in any way, since there it is a question of getting ready for something that is a forerunner (of things to come) (*HB*, p. 329).

Heidegger's critique of Sartre's reading of *Being and Time* is that Sartre saw it as a work of metaphysics, which *Being and Time*, in fact, attempted to unsettle, deconstruct and "destroy."[27] What Sartre purports to find in Heidegger on "the relationships between *essentia* and *existentia*" does not have a place in Heidegger's fundamental ontology, which is all about searching for the ground of metaphysics in the ontology of existence [*Da-sein*].

A few pages later in the "Letter," where he refers to Sartre's lecture, Heidegger questions whether his way of thinking can be construed as a kind of humanism, which, by associating Heidegger with existentialism, Sartre has intimated. His reply is unambiguous:

> Surely not, insofar as humanism thinks metaphysically. Surely not, if it is existentialism and it is represented in the way Sartre articulates it in this sentence: *précisément nous sommes sur un plan où il y a seulment des hommes* (*L'Existentialisme est un Humanisme*, p. 36). As thought about in *Being and Time*, this would read: *précisément nous sommes sur un plan où il y a principalement l'Être*. But where does *le plan* [the plane] come from, and what is *le plan*? *L'Être et le plan* are the same thing [*dasselbe*]. In *Being and Time* (*GA* 2, p. 281), it is said, intentionally and cautiously: *il y a l'Être*: "there is [*es gibt*]" be[-ing] [*Sein*]. But "*il y a*" inaccurately translates "*es gibt* [there is]." For the "it," which in this case "gives," is be[-ing] itself. The (word) "gives," however, names the essence [*Wesen*] of be[-ing] (as) giving [*gebende*], (i.e.,) granting [*gewährende*], its truth. Giving itself [*das Sichgeben*] in (the sphere) of what is open [*das Offene*] (along) with the latter itself is that very same *be*[-ing] (*HB*, p. 334).[28]

For Heidegger, human reality takes place on a "plane" dominated by be[-ing], not human beings. The difficulty, as Heidegger sees it, lies in understanding the way be[-ing] "is" *not*. Transposing this idea to human beings, Sartre has said that human beings precisely *are not*. But that is not what concerns Heidegger in *Being and Time*, or anywhere else in his writings. The passage from Sartre's lecture discussed by Heidegger points to where an error in Sartre's reading of *Being and Time* has occurred that led to a misunderstanding of Heidegger's thought.

Later, Heidegger also cites Sartre "(so far)" (and Husserl) for failing to "recognize the essentiality of what is historical about be[-ing]," which is one reason, Heidegger claims, why existentialism cannot "enter into that very dimension within which a productive discussion with Marxism would for the first time be possible" (*HB*, p. 340). This is a separate issue, of course, but Sartre may have taken Heidegger's words to heart when he abandoned the early aristocratic solipsism of *Being and Nothingness*, where any choice is possible, and later, as a student of Marx, embraced the importunity of social problems on the range of one's possible choices, recognized the impact of one's early environment on the scope of his possibilities, and, periodically leaving aside his meditations on literature, became a political activist.[29]

The substance of Heidegger's disavowal of existentialism is that it is an incarnation of metaphysical thinking in which human beings are the focal point of reference. By contrast, Heidegger's interest is be[ing].[30] Although, Sartre freely adopts some of the terminology of *Being and Time* in *Being and Nothingness,* we must conclude that Sartre's Heidegger is not one Heidegger himself would recognize. References to "being-in-the-world" and "human reality" [*Dasein*] notwithstanding, the upshot of Heidegger's thought is lost on Sartre, who sees Heidegger as continuing the tradition of Husserl, a philosopher who figures even more prominently in *Being and Nothingness* than Heidegger.

In my opinion, much of Sartre's confusion lay with his misunderstanding of Heidegger's concept of *Existenz* in *Being and Time*. Although *Existenz* is a fundamental notion in the early Heidegger, to which Sartre appeals, Sartre understands *Existenz* existentially and metaphysically, not ontologically, as Heidegger does.[31] In addition, like so many others, Sartre read *Being and Time* as an essay on philosophical anthropology, and understood dread [*Angst*] and care [*Sorge*], for example, as emotional dispositions rather than as ontological structures.

Perhaps more than anything else, however, Heidegger's seriousness attracted Sartre, who was being held as a prisoner of war, not sitting at a Left Bank café, when he immersed himself in Heidegger's *magnum opus*. The stupidities of World War II very likely attracted Sartre to Heidegger's discussion of the no-thing [*Nichts*] in his inaugural address, from 1929, "What Is Metaphysics?," although his circumstances and mood precluded Sartre from seeing that Heidegger's analysis of *das Nichts* in that text was about the co-valence of be[-ing] [*Sein*] and no-thing [*Nichts*], and not a meditation on post-Nietzschean nihilism.[32] In the end, we must never forget that the existential psychoanalysis of *Being and Nothingness* was conceived in the glare of a political prisoners camp's searchlight, which revealed the absurdities of war and death, as much as in light of Sartre's insights into the lack or absence at the heart of human reality.[33]

HEIDEGGER AND SARTRE'S EXISTENTIAL PSYCHOANALYSIS

It remains for me to review whether and to what extent Heidegger's thought influenced Sartre's existential psychoanalysis. Heidegger influenced psychoanalysis, but not the Sartrean version. The path of his influence ran, not through France, but through Switzerland, first via Ludwig Binswanger's *Daseinsanalyse* [existential analysis], which Heidegger rejected, and then via the *Daseinsanalytik* [analytique of existence] of Medard Boss, with whom Heidegger collaborated for many years.[34] Nor did Heidegger's ontology have an impact on Sartre, since Sartre failed to understand Heidegger's fundamental ontology in *Being and Time*. Whereas be[-ing] (in relation to temporality) is the central question of Heidegger's fundamental ontology, the human being is at the heart of Sartre's phenomenological ontology. Sartre's misinterpretation of *Being and Time*, which is in the immediate background of his existential psychoanalysis, may be traced to his misunderstanding of what Heidegger means by *Existenz*. I conclude with a few comments about this concept.

In his "Letter on 'Humanism'," Heidegger quotes the well-known sentence from *Being and Time* on which Sartre ran afoul: "Das 'Wesen' des Daseins liegt in seiner Existenz." He is quite clear that "the main point of 'Existentialism' has not the least bit in common with the sentence." What *does* the sentence mean? "The 'essence' of existence is in its way of life."[35] The word *Wesen* is given in scare quotes, indicating that Heidegger is not using the word in its standard acceptation, which refers to a thing's nature, *essentia* or *Essenz*, which Sartre confounds with the thing's coming to pass [*Wesen*] or, as Heidegger will say, the remaining or lingering of the thing, in this case existence [*Dasein*].

To understand what Heidegger means by the *Wesen* of something, we must look at the meaning of the verb *wesen*, of which *Wesen* is the substantive form. In the lecture course *Was heißt Denken?* [*What Calls for Thinking?* or *What Is Thinking?*], Heidegger's first course after having been permitted to return to teach after a politically imposed exile, he explains (Heidegger, 1954, p. 143):

> Used verbally, the word *wesen* is (found in) the old German word *wesan*. It is the same word as *währen* [to hold out or endure] and means to remain [*bleiben*]. *Wesan* stems from the Sanskrit [*altindisch*] (word) *vásati*, which means he lives (at) [*wohnt*], he stays (at) [*weilt*]. What is inhabited [*das Bewohnt*] is called the household [*Hauswesen*]. The verb *wesan* speaks of [*besagt*] (an) enduring staying on [*bleibendes Weilen*].

As Heidegger makes abundantly clear in the "Letter," what he had in mind, in 1927, in the sentence that helped launch French existentialism was not a distinction between man's *essence* or nature and his *existence*. Heidegger explains that when Sartre says the fundamental belief of existentialism is that "a way of life [*Existenz*] precedes what is by nature [*Essenz*]," he "takes (the terms) *existentia* and *essentia* [correlatively, in French, *existence* and *essence*] in the sense they have for metaphysics, which since Plato has said that essentia [as possibility] precedes existentia [as actuality]."[36] Whether it is also the case that "Sartre reverses this proposition" remains to be decided, but whether or not Sartre upends Platonic metaphysics, the undertaking and its result are still metaphysics. Sartre failed to understand Heidegger's sentence, which says that the coming to pass [*Wesen*] of existence [*Dasein*] is its way of life [*Existenz*] at any given moment. For that reason, he sees Heidegger as an existentialist, rather than someone whose sole interest lay in elucidating the be[-ing] of human reality, not an individual's way of living his life, which was a central concern of many Occupation-traumatized French intellectuals of Sartre's generation.

We recall that the Club Maintenant, to whose members and guests Sartre gave his lecture on whether existentialism is a humanism, was comprised of a group of intellectuals whose manifesto was saturated with "*l'importance exceptionelle d'aujourdui*," with the extraordinary relevance of the now, of what is going on "today"—in that instance, in fact, on Monday, October 28, 1945.[37] It was a group "*attentifs à 'maintenant'*," engaged, caught up in "the moment." The study of history could wait.[38] Perhaps this prompted Heidegger's comment in the "Letter," cited earlier (Heidegger, 1967, p. 340), on the "historicity" of at-homeness, that is, of be[-ing]. To repeat, Heidegger was concerned about the be[-ing] of any way of life, not about a particular way of life, whether existential or parochial, sacred or profane, philosophical or materialistic. For Sartre, of course, the primary goal of existential psychoanalysis is to reveal the project that sets in motion and sustains a given way of life.

Sartre's haunting question remains: *Am* I ever? It is a question of enduring interest to existential psychoanalysis, whether of Sartrean or Heideggerian inspiration. In keeping with one of the purposes of this chapter, in closing I will try to respond to the question as I think a Sartrean existential psychoanalyst might. The response will take the form of a brief case study.

A playwright who is "stuck" and unable to write visits an empirical psychoanalyst. For such an analyst, the analysand *has* writer's block. The writer who cannot write is inhibited. Her inhibition is *explained* by recovering memories of early childhood events that produced a conflict which has now returned in the form of the symptom "writer's block." The symptom is understood to be symbolic of a once actual, now long-forgotten conflict.[39] In other words, current stressors have revived a latent conflict, which has broken through the barrier of repression and now dominates the consciousness of the patient.[40] Treatment consists of interpreting the recollected memories and related contemporary events, such as dreams and mistakes (parapraxes), and reconstructing the patient's past. Only knowledge of the patient's distant past provides the analyst and analysand with what they need to know, in order to overcome the analysand's inhibition.

A Sartrean existential psychoanalyst takes an entirely different perspective on the "complaint."[41] He sees the not-writing as embodying the analysand's existence at that moment in her life. She is understood *as* her inability to write. She *is* this incapacity. She does not *have* something ("writer's block"); nor is she missing something (writing ability). For Sartre, existence itself is a lack, the nothingness out of which human being-for-itself fashions a life, without ever completing the job, of course, and without the human being ever becoming something fixed, once and for all. The symptom is a manifestation of something present, the individual's fundamental project, which must be revealed and understood. In this case, the project of being-a-writer must be articulated as it came to life for the analysand.

The analysand talks about her life. As might be expected, much of the discussion focuses on writing, playwrights, plays and criticism, but the moment of therapeutic consequence occurs when the analysand revisits her decision *to be a writer*. She might say, for example: "By writing I was existing. I was escaping from the grown-ups."[42] Her current not-writing is a mode of *being a writer* and must be understood as such.

Where does the therapist go from here? The results of treatment by empirical psychoanalysis are reported to be less than impressive. For Freud, psychoanalysis was successful when the analysand was again able to work and to love; for Jung, analysis energized the on-going process of individuation. But what is the therapeutic benefit of Sartrean existential psychoanalysis, if its goal is to articulate the analysand's primordial project?

To have revisited one's project is to have made contact with the wellspring of one's freedom, and that in itself is therapeutic, in the original sense of therapeutics as attending to what is most essential. Existential psychoanalysis has no interest in adjusting or altering behavior. It shares with empirical psychoanalysis an appreciation for insight, not for purposes of self-transformation, however, but as the way in to the wellspring of one's freedom and the responsibility for one's life that is part of our being freedom.

EPILOGUE: CHANGE

A few comments about change in the context of Sartrean psychoanalysis may be in order. They arise from considering an element of the Heideggerian perspective on clinical practice that eluded Sartre and, as such, may shed a bit of light on what the intellectual encounter between the two philosophers might have led to, if Sartre had understood Heidegger's thought as he developed his existential psychoanalysis.

When the individual mentioned above was no longer writing, we would say she had changed. For Sartre, however, change is not possible, since each of us is working out the same unifying project embarked on at that critical moment of fundamental decision. We are an uncompleted project—the same project. We do not change, because our project remains the same. On the other hand, for Heidegger, change is possible—thanks to be[-ing], of course, and not to the will or actions of the human being in question. Is a rapprochement between Heideggerian and Sartrean insights possible on the matter of change? Yes, and I think it might be approached in the following, preliminary way.

We are always changing and therefore never are what we are (as Sartre would say, but from a different perspective and for different reasons), yet all change is caught up in the structure of existence [*Da-sein*], which is that site where be[-ing] comes about (as Heidegger would say).[43] It happens, however, that we notice and pay attention only to dramatic changes in the course of everyday life.[44] One who has thought of herself only as a writer is noticeably changed when she can no longer write. My point is that Sartrean analysis seems to be concerned with such dramatic changes, while also claiming that such change cannot occur. The contradiction disappears, however, if we admit that change continually occurs, not in existence, however, but in be[-ing]. Heidegger's fundamental ontology allows for this, although Sartre's phenomenological ontology does not.[45]

Our not-writing individual seems to confirm my observation, that one *is* anything in particular only when she is actually engaged in or practising what the designation indicates we should expect to see her doing. Is this merely a trivial observation? I think it is not only not trivial, but fundamental to what distinguishes existential psychoanalysis from empirical psychoanalysis, in that it points to what Heidegger considered to be the essential feature of existence [*Da-sein*]; namely, that it is always a way of living one's life [*Existenz*]. As I read the famous sentence: "The coming to pass of existence lies in its way of life." In practice, then, existential psychoanalysis is not as far from Heideggerian *Daseinsanalytik* as it might appear to be at first blush. The difference between the two forms of analysis is found in Sartre's emphasis on choice (which Heidegger claims is never ours), while Heidegger attributes all change to a dispensation of the fundamental at-homeness or be[-ing] of existence.

NOTES

1. "Existential Psychoanalysis," which is the first section of Chapter 2 ("Doing and Having") of Part 3 ("Being-for-Others") of *Being and Nothingness* (the other two sections are "The Meaning of 'to Make' and 'to Have': Possession" and "Quality as a Revelation of Being"), was published, in 1953, as the first half of a book entitled *Existential Psychoanalysis* along with another section of *Being and Nothingness*, "Bad Faith," which is Chapter Two of Part One ("The Problem of Nothingness") of the work. That same year, the rest of the translation of *L'Etre et le Néant* was published, but without these sections, in a "special abridged edition." In the edition of *Existential Psychoanalysis*

published by Philosophical Library, there is an introduction by the translator, Hazel Barnes. Henry Regnery also published an edition of *Existential Analysis* in 1953. Beginning in 1962, it contained an introduction by Rollo May, quite to the surprise of the translator, whose introduction was omitted. Hazel E. Barnes (personal communication, March 18, 2002). A complete text of *Being and Nothingness* first appeared only in 1956. A psychoanalyst and one of Sartre's keenest critics, François Lapointe (Lapointe, 1971, p. 17) notes that "the fourth and final part of *Being and Nothingness* [of which "Existential Psychoanalysis" is the core] may be seen as one long dialogue with Freud" or what Sartre will call "empirical psychoanalysis." For the saga of the translation and publication of *Being and Nothingness*, see Barnes (1997, pp. 153–155). *Existentialism and Humanism* (Sartre, 1947) was an English translation of Sartre's 1945 lecture "L'Existentialisme est un Humanisme."

2. While Erikson's psychosocial approach is empirical, he stresses existential turning points in a life history more than many traditional psychoanalytic writers. In his existential psychoanalysis of his contemporary, Jean Genet (Sartre, 1963a, 1963b, 1963c), Sartre treats his subject as an historical figure, as he does himself in his self-analysis (Sartre, 1964). As reported by Clément (1990, p. 57), Sartre once hinted that he had spent a few sessions "on the couch" with a psychoanalyst, presumably of the traditional variety. On Sartre's self-analysis, see Fell's (1968) essay "Sartre's *Words:* An Existential Self-Analysis." This article prompted a response by the psychiatrist Keen (1971), who argues for a "rapprochement" between Freudian psychoanalysis and Sartre's existential psychoanalysis.
3. The influence of Heidegger is already evident in *The Emotions* and *The Psychology of Imagination*, both of which were published just at the time of Sartre's immersion in 1939–1940 in Heidegger's *Sein und Zeit.*
4. On Sartre's critique of Freud, see the excellent article by Phillips (1986).
5. See, for example, *Critique of Dialectical Reason* (Sartre, 1976, pp. 17–18).
6. One commentator, Schrift (1987), sees continuity between Sartre's early (phenomenological–ontological) and later (dialectical) versions of existential psychoanalysis, arguing that the *ontological approach* to the individual, the for-itself, project, facticity and lack in the early Sartre of *Being and Nothingness* finds parallel and equivalent expression in the *structural* approach to the social in terms of totalisation, scarcity, praxis and need Sartre worked out the *Critique de la Raision Dialectique*. Sartre confirms Schrift's observation in a note to that work (Sartre, 1960, pp. 285–286).
7. Cannon (1999, p. 25) goes so far as to declare that "the screenplay may now be judged to contain some of Sartre's finest writings."
8. Quotations are from the 1956 edition of *Being and Nothingness* (*BN*). The translator notes that the sections published in *Existential Psychoanalysis* were revised for the first complete edition of *Being and Nothingness*.
9. *BN*, p. 598.
10. It is noteworthy that, for some schizophrenics, human beings lose their human qualities, in part or as a whole, and are experienced as non-human things.
11. The reversal in order of presentation of the two sections in *Existential Psychoanalysis* may have complicated the early reader's encounter with Sartre's existential psychoanalysis, since what he recommends in the section by that name presumes an understanding of what we learn in the section on "Bad Faith." The translator/editor, Hazel Barnes, justified the arrangement by arguing that "'Bad Faith' in actuality contains criticism and in a sense the specific application of ideas more fully developed in the other essay" ("Translator's Preface," p. 7).
12. Yalom (1980, p. 222) nicely summarizes Sartre's own project and what might be called Sartre's therapeutic stance: "Sartre considered it his project to liberate individuals from bad faith and to help them assume responsibility."
13. The radical freedom Sartre has in mind has been criticized as naive and disingenuous. Sartre himself modified the absoluteness of this freedom later in his life, when he admitted that early social conditioning affects the range of possibilities from which one makes his choices.

14. Two authoritative sources of this view are Erikson's (1994) *Identity, Youth and Crisis* and Blos's (1985) On *Adolescence.* Among other issues, the question of change looms large in a critique of developmentalism in psychology that any serious discussion of the view would entail. See my comments on change in Sartrean and Heideggerian analysis in the concluding sections of this chapter.
15. For empirical psychologists, the shibboleths of real science are quantifiability, measurability and predictability.
16. de Beauvoir (1948), *Pour une Morale de l'Ambiguité.*
17. The first part of Light's (1987, pp. 3–39) little volume *Shuzo Kuki and Jean-Paul Sartre. Influence and Counter-Influence in the Early History of Existential Phenomenology* contains most of the relevant references to Sartre's encounter with Heidegger's thought. Evidently, the two philosophers met in person only once, on December 23, 1952, for about an hour and a half in Freiburg, where Heidegger lived. For details, see Contat and Rybalka (1974, p. 18) and Petzet (1993, p. 81). Ott (1993, p. 329) reports that a meeting between Heidegger and Sartre had been scheduled for sometime in late 1945, but did not materialize. On December 10, 1940, while a prisoner of war, Sartre (Sartre, 1992, p. 301) wrote to his "Beaver," Simone de Beauvoir: "Je lit Heidegger et je ne me suis jamais senti aussi libre (I am reading Heidegger and I have never felt so free)." According to Sartre's bibliographers (Contat & Rybalko, 1974, pp. 10–11), he gave a course on Heidegger to a group of priests while he was a prisoner of war. Sartre's first sustained reading of *Being and Time* had begun in 1939, although he had tried to read Heidegger's book as early as 1933 (Light, 1987, p. 24 and pp. 6 and 78). *Sein und Zeit* was not published in French until 1964 in a translation by Rudolf Boehm and Alphonse de Waelhens. On Heidegger's initial response to Sartre's reading of his works, see Safranski (1998, p. 321), where a letter from Heidegger to Sartre, dated October 28, 1945, is quoted: "Here for the first time I encountered an independent thinker who, from the foundation up, has experienced the area out of which I think. Your work shows such an immediate comprehension of my philosophy as I have never before encountered."
18. Heidegger's correspondence with Beaufret had begun at least a year earlier with a letter dated November 23, 1945. There Heidegger refers to two articles on existentialism by Beaufret in issues 2 and 5 of the journal *Confluences* (March and June/July 1945). In fact, Beaufret (1945a, 1945b, 1945c, 1945d, 1945e, 1945f) published in the journal a series of six articles "A Propos de l'Existentialismse." The fifth article announced at its conclusion that the series would end with the next installment (Beaufret, 1945e), yet the title of the sixth article (Beaufret, 1945f) suggests that the reader could expect a second part on "Existentialism and Marxism," which evidently never appeared in the journal. Lapointe (1981, p. 316) confirms this in his entry on the series of articles by Beaufret. I give these details, in order to raise the question of how much of what Heidegger learned about French existentialism was acquired by reading Beaufret's articles—on Kierkegaard's influence on the movement (Beaufret, 1945a), and especially the first part of Beaufret's discussion of Sartre (Beaufret, 1945d)—and how much through direct contact with Sartre's works. Of course, Heidegger knew Sartre's lecture, which was given on October 25, 1945, but published only in February 1946, since he quotes it in his "Letter on 'Humanism'," but it is less clear how much Heidegger knew of, for example, *L'Etre et le Néant*, which was written in the early 1940s and published in 1943 (Contat & Rybalko, 1974, pp. 82–84). In the first of the articles that Heidegger had read when he wrote to Beaufret in late 1945, Beaufret cites *Sein und Zeit* as the source of "une analytique de la condition humaine." Referring to Heidegger with great admiration, he writes: "Imaginons un Aristote qui, par la force de la logique qu-il s'impose dans ses recherches, en viendrait à ne traiter que les problèmes pathétiquement débattus par Pascal et Kierkegaard" (Beaufret, March 1945a, p. 197): "Imagine an Aristotle with the power of logic at his disposal that he applied to his areas of study, who has set out to deal with the extremely moving matters discussed by Pascal and Kierkegaard." The remaining pages of Beaufret (1945a) are chiefly about Heidegger, who may not have seen the works (Beaufret, 1945b, 1945c), that are entirely devoted to his thought. Beaufret (1945d, p. 532) points out Sartre's debt to Heidegger, who is mentioned by name only once more in the article (p. 534). This little-known series of articles by Beaufret

may play an important part in the story of how Heidegger came to know about Sartre's work. More important for the present discussion, Beaufret points out Sartre's misinterpretation of Heidegger's notion of *Existenz* and his misreading of the sense of Heidegger's statement that the "essence of human reality lies in its existence," which Sartre transforms into the observation that in human beings, "essence precedes existence." Beaufret also stresses Heidegger's emphasis on the future, which contrasts with Sartre's "existentialist" emphasis on the present moment (Beaufret, 1945a, p. 416). In a stunning twist, Beaufret concludes with the remarkable observation that Heidegger's philosophy is "le platonisme de notre temps" (Beaufret, 1945c, p. 422)—"the Platonism of our time." It is important to recall that the correspondence is taking place at the beginning of Heidegger's post-war exile from teaching (1945–1951). Safranski (1998, pp. 348–350) recounts some of the circumstances of the period.

19. A slightly reworked version of the lecture was published as *L'Existentialisme est un Humanisme* (Sartre, 1947). The following year it was translated into German (no translator named) with the title *Ist der Existeniaslismus ein Humanismus?* and into English, with a title that does not hint at Sartre's question about the relation between existentialism and humanism: *Existentialism and Humanism.* In fact, in the lecture, Sartre asks whether existentialism is a humanism. He does not make the assertion. As his bibliographers point out (Contat & Rybalko 1974, p. 133), in the text itself, Sartre himself refers to the lecture as "L'Existentialisme est-il un Humanisme?," "[a]lthough the lecture was announced in the papers as 'Existentialism Is a Humanism'." A second English translation, by Philip Mairet (to which I will refer: *EH*), was published in 1948. The lecture is complemented by a protocol of discussions with "opponents of his teaching" (prefatory "Note"), most notably the Marxist, Pierre Naville (*EH*, pp. 57–70). Sartre's bibliographers (Contat & Rybalko, 1974, p. 13) aptly note: "It is worth recalling, however, that the work constitutes a rather poor introduction to Sartre's philosophy, above all for readers who have not been forewarned. Focused primarily on ethical questions, it popularizes the outstanding claims of existentialism at the price of making a sort of moralistic travesty out of them. This is, moreover, the only work Sartre has largely rejected." Was Heidegger "forewarned" or is he responding primarily to the Sartre of *L'Etre et le Néant?* Contat and Rybalko (p. 132) add: "This lecture marks an unforgettable day in the anecdotal history of existentialism: so many people came that women fainted and the lecturer could scarcely make himself understood."
20. In connection with our "abandonment," Sartre refers to Heidegger, for whom, he says, this is "a favorite word" (*EH*, p. 32). Here Sartre is probably translating Heidegger's *Geworfenheit* as used in *Sein und Zeit.* The reference to Heidegger at this point may suggest to the reader that Heidegger is an ethicist, which a close reading of his writings does not confirm. This has not prevented some scholars from claiming to have found the basis for an ethics in Heidegger's thought. See, for example, Hatab (2000).
21. We should not forget that what, for Sartre, seem to be affective reactions (anguish, abandonment, and despair) are, for Heidegger, ontological features of human be-ing [*Da-sein*]. The influence of Heidegger's *Sein und Zeit* is unmistakable in this section of the lecture. One thinks of the passage in *Being and Time* (1996, p. 34), "Higher than actuality stands possibility."
22. There are some difficulties with these principles, but a critique of (1) existentialism and truth, (2) the status of existentialism as a theory or an ideology, and (3) how *la condition humaine* differs from human nature would require another series of studies.
23. At this point, Sartre disassociates himself from classical humanism and allies existentialism with the "fundamental" meaning of humanism, which he gives in the next few sentences of the lecture. One source of Sartre's affinity for Heidegger may be traced to their shared distrust of value-based philosophy. As we will see later on, however, Heidegger may not have fully appreciated Sartre's distinction at this point in the lecture.
24. By his own admission, Sartre's reading of Heidegger had been an arduous ordeal. Even German-speaking readers had complained of the difficulty understanding Heidegger's thought as expressed in *Sein und Zeit.*

25. As we have seen, the "Brief über den 'Humanismus'" was Heidegger's response to a letter from Jean Beaufret. It was first published in 1947 along with Heidegger's 1940 lecture "Platons Lehre von der Wahrheit" as a separate booklet, which has been reprinted many times. A French translation by Roger Munier was published in two installments in the periodical *Cahiers du Sud* **40** (pp. 319–320) only in 1953. In April 1949, the British periodical *World Review* (#2, New Series, pp. 29–33) published an anonymous translation of a few excerpts of the text. This was apparently the *first* Heidegger published in English anywhere. In 1962, a full English translation, by Frank Lohner, appeared and has been reprinted in several volumes since then. Two other English translations have been published. I refer to Volume 9, *Wegmarken*, of the *Gesamtausgabe* edition, Klostermann, Frankfurt [*HB*] of Heidegger's writings. All translations are my own.
26. The equation of *essentia* with possibility and *existentia* with actuality had been made by Heidegger earlier in the text. In fact, however, Sartre has not done what Heidegger claims he has. Sartre says we "remain in the realm of possibilities," which has priority over actuality. He understands actuality as being. My rendering of *Existenz* with "life" or, better, "a way of life" is the result of a long struggle with Heidegger's thought and *Being and Time's* language, which Heidegger himself eventually dismissed as inadequate to his way of thinking. For a full treatment of what is at stake here, see my *Translating Heidegger* (Groth, 2004). Briefly, the problematic is as follows. For Heidegger, the outstanding feature of existence [*Da-sein*] (the human way of be-ing [*Seiende*]), is *Existenz*, a distinctive way of living one's life [*Leben*]. The emphasis on our always existing as a particular way of life is the "existential" element of Heidegger's thought that attracted (among other French intellectuals) Sartre, but which, I will argue, Sartre did not fully grasp. His misunderstanding of Heidegger's thought depended on his misreading of Heidegger's usage of *Existenz* in *Sein und Zeit* and other early texts.
27. Heidegger's term *Destruktion* is perhaps better rendered with de-structuring or disassembling.
28. Sartre's words in the passage quoted may be likened to the English expression "plane of existence." However, the French word *plan* also corresponds to Heidegger's term *Entwurf* (project). Earlier on in the lecture, Sartre had said: "Man is at the start a plan which is aware of itself...; nothing exists prior to this plan; there is nothing in heaven; man will be what he will have planned to be. Not what he will want to be." One English translation confuses the two sense of the French word *plan*.
29. Sartre (1974b, pp. 199–223) includes the author's late views on psychoanalysis.
30. It is interesting to note that many existentialists underwent major transformations of their *Existenz*, while Heidegger remained absorbed by one question throughout his life—the question about the meaning of *Sein*. Among the Christian existentialists, Karl Jaspers *is*, by turns, psychiatrist and philosopher. Gabriel Marcel *is*, by turns, playwright, theologian and philosopher. The major atheistic existentialist, Sartre himself, *is* variously a novelist, playwright, philosopher, existential psychoanalyst, and political activist. By contrast, the inveterate interrogator of the *Seinsfrage* [question about be[-ing]], Heidegger *is* nothing at all. Eventually even disowning the designation philosopher, he admits only to thinking [*Denken*], which is not to be confused with cognitive activity but is construed as a form of thanking [*Danken*]. See Heidegger (1968, p. 138), which he is said to have referred to as his "favorite book" among those he published.
31. I will return to this issue at the conclusion of the chapter.
32. Heidegger (1976, pp. 103–122). After 1943, the text was published with an important "Postscript" (1976, pp. 303–312) and, in editions from 1949 on, the lecture was preceded by an equally important "Introduction" (1976, pp. 365–384). Heidegger's lecture was first published in 1929, with a French translation by Henri Corbin and others available only in 1938.
33. Once again, we should not forget that, at the beginning of the Second World War, for many French existentialists, including Sartre, Heidegger's philosophy was tainted by his involvement with National Socialism in 1933 and for many years following.
34. Several of Binswanger's essays were published (Binswanger, 1963) under the obviously Heideggerian title *Being-in-the-World*. The book contains essays from 1930 on. Heidegger's influence on Boss is evident beginning with his *Psychoanalysis and Daseinsanalysis* (Boss, 1963) and

concluding with the posthumously published *Zollikon Seminars* (Heidegger, 2001), which contains Heidegger's critique of Binswanger's *Daseinsanalyse*. The book is a collection of protocols of seminars Heidegger held for residents in psychiatry being supervised by Boss. The seminars were held at Boss's home in the Zollikon district of Zurich. *Zollikon Seminars* also includes correspondence and accounts of conversations between Boss and Heidegger from the years 1947 to 1972. In the course of the seminars, Heidegger also dismissed the work of Erwin Straus, Wolfgang Blankenburg and Jacques Lacan, whom he diagnosed as perhaps being in need of a psychiatrist himself. See my review (Groth, 2002, pp. 164–166) of *Zollikon Seminars*. Unlike Sartre's existential psychoanalysis, Boss's *Daseinsanalytik* was not caught up in the metaphysical *Weltanschauung* inherited from the post-Socratic Greek philosophers. We know that Heidegger rejected Freud's psychoanalysis and, like Sartre, saw it as a form of empiricism. In an interview with the author, in June, 1976, Boss recounted observing that, after having read a few of Freud's texts, Heidegger asserted that he could not accept Freud's approach. On Heidegger's opinion of Freud, see Groth (1976, p. 69, no. 1).

35. In the standard translation by John Macquarrie and Edward Robinson (Heidegger, 1962, p. 67), the sentence runs: "The 'essence' of Dasein lies in its existence." Joan Stambaugh's (Heidegger, 1996, p. 39) rendering is more in the spirit of Heidegger's thought: "The 'essence' of the being lies in its to be."
36. Heidegger had established these equivalences—*essentia*/possibility, *existentia*/actuality—in the preceding pages of the "Letter."
37. The preceding and following quotations are from the "Avant-propos" to Wahl's (1947, pp. 7–9) *Petite Histoire de "L'Existentialisme,"* which never made it into the English translation, *A Short History of Existentialism* (Wahl, 1949). The preface is signed, collectively, "C.M." The translation also omits Wahl's remarkable reflections on "Kafka and Kierkegaard," which conclude the French edition. Wahl's little volume was published by the same house as *Existentialism and Humanism*. On her experiences with the publisher, see Hazel Barnes' memoir (1997, pp. 149–150, 153, 159).
38. On that very day, as it happens, Heidegger had written a letter to Sartre touting his "comprehension of my philosophy as I have never before encountered." The letter was uncovered by Frédéric de Towarnicki and published for the first time in the early 1990s. See Safranski (1998, pp. 349 and 448, no. 27) for details and a reference (p. 350) to a document in which Heidegger credits Sartre's understanding of *Sein und Zeit* as decisive for Heidegger's later reception among French intellectuals. Safranski notes that around the same time the Club Maintenant was meeting to listen to Sartre's lecture on existentialism and humanism, Heidegger was in his Alpine lodge, near Todnauberg in the Black Forest of southwest Germany, considering whether to write a philosophical analysis of skiing and pondering the denazification proceedings he was facing.
39. Repression indicates that the individual has forgotten uncomprehended or emotionally overwhelming events but has *also* forgotten that something once known has been forgotten.
40. The old conflict may be compared to a microorganism that has lain inactive since it manifested as a disease (chicken pox). Under the influence of stressors that weaken repression (the immune system), the organism revives in a somewhat altered form as the current symptom (shingles).
41. The best sources of information about Sartrean clinical practice are by Cannon (1991, 1999). An earlier version of Cannon (1999) was given at the Wagner College Conference on Existential Psychotherapy, which was held on May 1, 1999, hosted by the author.
42. The quotation is, of course, from Sartre's own *Words*.
43. Heidegger's verb here is *west*, the third-person singular of *wesen*. See the passage from *Was heißt Denken?*, cited above. As it happens, the passage was omitted from the English translation.
44. Something similar can be said about moods. Each of us is always in a mood, but we are usually not aware of the mood unless it is especially strong or is markedly different from the mood we were in that preceded it.
45. When Sartre says that we never are what we are and, at the same time, always are what we are not, Heidegger agrees but takes us a step further ontologically, by suggesting that we have nothing to do with such change. This is not a sort of crypto-determinism or fatalism. Heidegger simply wants to say that change is of be[-ing] not by existence.

REFERENCES

Barbone, S. (1994). Nothingness and Sartre's fundamental project. *Philosophy Today*, *38*(2), 191–203.

Barnes, H. (1997). *The story I tell myself. A venture in existentialist autobiography*. Chicago: University of Chicago Press.

Barnes, H. (2002). Letter dated March 18, 2002.

Beaufret, J. (1945a, March). A Propos de l'Existentialisme (I). *Confluences*, *2*, 192–199.

Beaufret, J. (1945b, April). A Propos de l'Existentialisme II. Martin Heidegger. *Confluences*, *3*, 307—314.

Beaufret, J. (1945c, May). A Propos de l'Existentialisme (III). *Confluences*, *4*, 415–422. (which is a continuation of the discussion of Heidegger)

Beaufret, J. (1945d, June–July). A Propos de l'Existentialisme (III) [Suite]: Jean-Paul Sartre. *Confluences*, *5*, 531–538.

Beaufret, J. (1945e, August). A Propos de l'Existentialisme (III): Jean-Paul Sartre (Suite). *Confluences*, *6*, 637–642.

Beaufret, J. (1945f, September). A Propos de l'Existentialisme III. Conclusion: Existentislisme et Marxisme (1). *Confluences*, *7*, 764–771.

Benda, C. (1966). What is existential psychiatry? *American Journal of Psychiatry*, *123*(3), 288–296.

Binswanger, L. (1963). *Being-in-the-world*. New York: Basic Books.

Blos, P. (1985). *On adolescence. A psychoanalytic interpretation*. Glencoe: Free Press.

Boss, M. (1963). *Psychoanalysis and daseinsanalysis*. New York: Basic Books. (First published in 1957)

Canfield, F., & La Fave, L. (1971). Existential psychoanalysis. *The University of Windsor Review*, *7*(1), 30–36.

Cannon, B. (1991). *Sartre and psychoanalysis. An existentialist challenge to clinical metatheory*. Lawrence: University of Kansas Press.

Cannon, B. (1999). Sartre and existential psychoanalysis. *The Humanistic Psychologist*, *27*(1), 23–50. (This text is the best current introduction to Sartre's lasting and continuing contribution to clinical psychology and psychotherapy.)

Chessick, R. (1984). Sartre and Freud. *American Journal of Psychotherapy*, *38*, 229–238.

Clément, C. (1990). Contre, tout contre la psychanalyse. *Magazine Littéraire: Sartre dans tous ses Écrits*, *282*, 55–57.

Contat, M., & M. Rybalka. (1974). *The writings of Jean-Paul Sartre* (Vol. 1). Evanston: Northwestern University Press.

de Beauvoir, S. (1948). *The ethics of ambiguity*. Seacaucus: Citadel Press. (First published in 1947)

Erikson, E. (1958). *Young man Luther*. New York: W.W. Norton.

Erikson, E. (1969). *Ghandi's truth*. New York: W.W. Norton.

Erikson, E. (1994). *Identity, youth and crisis*. New York: W.W. Norton.

Fell, J. (1968). Sartre's words: An existential self-analysis. *Psychoanalytic Review*, *55*, 426–441.

Fell, J. (1979). *Sartre and Heidegger. An essay on being and place*. New York: Columbia University Press.

Groth, M. (1976). Interpretation for Freud and Heidegger: parataxis and disclosure. *International Review of Psycho-Analysis*, *9*, 67–74.

Groth, M. (2002). Review of Martin Heidegger, *Zollikon seminars*. *Journal of the Society for Existential Analysis*, *13*, *1*, 164–166.

Groth, M. (2004). *Translating Heidegger*. Amherst: Humanity Books.

Hanly, C. (1979). *Existentialsim and psychoanalysis*. New York: International Universities Press.

Hatab, L. (2000). *Ethics and finitude. Heideggerian contributions to moral philosophy*. New York: Rowman and Littlefield.

Heidegger, M. (1927). *Sein und Zeit*. Tübingen: Niemeyer.

Heidegger, M. (1954). *Was heißt Denken?* Tübingen: Niemeyer.

Heidegger, M. (1962). *Being and time* (J. Macquarrie & E. Robinson, Trans.). London: SCM Press.

Heidegger, M. (1968). *What is called thinking*? New York: Harper and Row. (First published in 1954)
Heidegger, M. (1976). *Wegmarken*. Frankfurt: Klostermann.
Heidegger, M. (1996). *Being and time. A translation of Sein und Zeit* (J. Stambaugh, Trans.). Albany: SUNY Press.
Heidegger, M. (2001). *Zollikon seminars. Protocols—conversations—letters*. Evanston: Northwestern University Press.
Keen, E. (1971). A rapprochement in the psychologies of Freud and Sartre. *Psychoanalytic Review, 58*(2), 183–188.
Lapointe, F. (1971). Phenomenology, psychoanalysis and the unconscious. *Journal of Phenomenological Psychology, 2*, 5–27. (This very important article suggests that Georges Politzer's early (1929) proposed re-envisioning of Freud, *Critiques des Fondements de la Psychologie*, Rieder, Paris, may lie at the source of Sartre's existential psychoanalysis.)
Lapointe, F. (1981). *Jean-Paul Sartre and his critics. An international bibliography (1938–1980)*. Bowling Green: Philosophy Documentation Center.
Light, S. (1987). *Shuzo Kuki and Jean-Paul Sartre. Influence and counter-influence in the early history of existential phenomenology*. Carbondale: Southern Illinois University Press.
Ott, H. (1993). *Martin Heidegger. A political life*. New York: Harper Collins.
Petzet, H. (1993). *Encounters and dialogues with Martin Heidegger 1926—1976*. Chicago: University of Chicago Press.
Phillips, J. (1986). Sartre and psychoanlysis. *Psychiatry, 49*, 158–168.
Phillips, J. (1988). Bad faith and psychopathology. *Journal of Phenomenological Psychology, 19*, 117–146.
Rumpel, H. (1995). Letter to the author, August 14, 1995.
Safranski, R. (1998). Martin Heidegger. *Between good and evil*. Cambridge: Harvard University Press.
Sartre, J.-P. (1936). *L'Imagination*. Paris: Alcan.
Sartre, J.-P. (1947). *Existentialism and humanism*. New York: Philosophical Library.
Sartre, J.-P. (1948a). *The psychology of imagination*. Seacaucus: Citadel. (First published in 1940. Sartre refers to an earlier study *L'Imgaination* (1936), which has not been translated)
Sartre, J.-P. (1948b). *The emotions. Outline of a theory*. New York: Philosophical Library. (First published in 1939)
Sartre, J.-P. (1953). *Existential psychoanalysis* (pp. 47–70). Chicago: Henry Regnery. (A later edition (1962) includes an introduction by Rollo May)
Sartre, J.-P. (1956). *Being and nothingness. An essay on phenomenological ontology*. Seacaucus: Citadel.
Sartre, J.-P. (1957). *The transcendence of the ego. An existentialist theory of consciousness*. New York: Farrar, Straus and Giroux. (First published in 1936)
Sartre, J.-P. (1960). *Critique de la raison dialectique*. Paris: Gallimard.
Sartre, J.-P. (1963a). *Saint Genet. Actor and martyr*. New York: George Braziller. (First published in 1952)
Sartre, J.-P. (1963b). *Search for a method*. New York: Knopf. (First published in 1960, this is the introductory essay for the *Critique of dialectical reason*)
Sartre, J.-P. (1963c). "Foreword" to R.D. Laing & D.G. Cooper, *Reason and violence. A decade of Sartre's philosophy 1950–1960* (pp. 6–7). New York: Pantheon Books. (Includes the original French text)
Sartre, J.-P. (1964). *The words*. New York: Braziller. (First published in 1964, this is Sartre's existential self-analysis)
Sartre, J.-P. (1967). *Baudelaire*, New York: New Directions Book. (First published in 1947, this is the first of Sartre's existential psychobiographies)
Sartre, J.-P. (1971). *The family idiot: Gustave Flaubert 18721–1857*. Chicago: University of Chicago Press. (First published in 1971)
Sartre, J.-P. (1974a). The Itinerary of a thought. In *Between existentialism and Marxism [Sartre on philosophy, politics, psychology, and the arts]* (pp. 33–64). New York: Pantheon Books. (First published in 1972)

Sartre, J.-P. (1974b). Psychoanalytic dialogue. In *Between existentialism and Marxism [Sartre on philosophy, politics, psychology, and the arts]* (pp. 206–223). New York: Pantheon Books. (First published in 1969)

Sartre, J.-P. (1974c). The man with the tape recorder. In *Between existentialism and Marxism [Sartre on philosophy, politics, psychology, and the arts]* (pp. 199–205). New York: Pantheon Books. (First published in 1969)

Sartre, J.-P. (1976). *Critique of dialectical reason* (Vol. 1). London: NLB. (First published in 1960. The second volume of the work, of which two chapters were completed, has not been published)

Sartre, J.-P. (1984). *War diaries: Notebooks from a phoney war, November 1939—March 1940.* New York: Verso. (First published in 1983)

Sartre, J.-P. (1985). *The Freud scenario.* Chicago: University of Chicago Press. (First published in 1984, the screenplay was revised without Sartre's editorial participation)

Sartre, J.-P. (1992). *Quiet moments in a war: The letters of Jean-Paul Sartre to Simone de Beauvoir, 1940—1963.* New York: Scribner. (First published in 1981)

Sartre, J.-P. (1993). *Witness to my life: The letters of Jean-Paul Sartre to Simone de Beauvoir 1926–1939.* New York: Scribner. (First published in 1981)

Schrader, G. (1959). Existential psychoanalysis and metaphysics. *The Review of Metaphysics, 13*(1959), 139–164.

Schrift, A. (1987). A question of method: Existential psychoanalysis and Sartre's *Critique of dialectical reason. Man and World, 20*(1987), 399–418.

Shouery, I. (1986). The paradoxical implications of the phenomenological reduction in Sartre's psychoanalysis. *Journal of Mind and Behavior, 7*(4), 585–590.

Spiegelberg, H. (1969). *The phenomenological movement* (Vol. 2, pp. 445–515 and 572). The Hague: Martinus Nijhoff (n. to p. 468)

Spiegelberg, H. (1972). *Phenomenology in psychology and psychiatry* (pp. 23–25). Evanston: Northwestern University Press.

Wahl, J. (1947). *Petite histoire de l'existentialisme.* Paris: Editions Club Maintenant.

Wahl, J. (1949). *A short history of existentialism.* New York: Philosophical Library. (First published in 1946. The French edition also contained a text by Wahl on "Kafka and Kierkegaard.")

Wilcox, R. (1975). *Jean-Paul Sartre. A bibliography of international criticism.* Edmonton: University of Alberta Press.

Yalom, I. (1980). *Existential psychotherapy.* New York: Basic Books.

CHAPTER 8

MEDARD BOSS' PHENOMENOLOGICALLY BASED PSYCHOPATHOLOGY

F. A. JENNER

PERSONAL BACKGROUND

Born on 4th October 1903 (he died in 1990), the son of a distinguished professional Swiss family in St Gallen, in Switzerland, Medard Boss became a student of medicine primarily in Zurich. There he was subsequently, in 1939, given the title of professor, but that was long after also studying in Paris, London, and Vienna. In Vienna, he had a training analysis with Sigmund Freud. Later in Zurich, for nearly a decade, there were monthly conversations with Carl Gustav Jung.

For various prolonged periods during the 1939–1945 war, Switzerland was surrounded by the Axis military forces, those of Germany, and Italy, and defeated (Vichy) France. A Nazi and Italian fascist invasion seemed a possibility. So Boss was, like many fellow Swiss citizens, conscripted into the army. During that time he had little medical work to do. His fortunately non-combatant healthy young fellow conscripts, all living together in the beautiful mountains, needed limited medical attention. He had, quite unusually for him, time on his hands, and he read in a newspaper about Martin Heidegger (1889–1976). As he was interested in the "nature of things" he obtained a copy of Heidegger's (1927/1962) *Sein und Zeit* (*Being and Time*). Not surprisingly his scientific medical training left him ill equipped—as most of us are—to understand much of what he read. Nevertheless he became somewhat mesmerized, and he kept going back to try a little harder. All that his philosophically informed acquaintances told him, and seemed to know, was that Heidegger was a Nazi. Ultimately Boss wrote to Heidegger, who replied by return of post and they developed a great mutual respect and friendship. It became apparent that Heidegger had always hoped to see some application of his philosophical ideas in medicine.

It was from Heideggerian philosophy, then, that Boss developed so much of his own "Daseinsanalysis," essentially as a therapeutic method. What he produced was an outlook based on a *menage à trois* of Heidegger's theories, and Freud's practices, while significantly adhering to the taxonomy of Kraepelin (1903).

Despite the fact that Boss had been told that Heidegger was a Nazi, once he got to know him he felt a strong need to defend the warm and considerate person he so venerated. He did feel that for a short while, Heidegger had made a very grave error of judgment of Hitler and of National Socialism. He noted however that the National Socialists significantly opposed Heidegger. They wanted Heidegger's work on Nietzsche changed, as it included Nietzsche's attack on anti-Semitism. Even so, there is a difference in attitude between Boss and Heidegger, Heidegger was an antidemocrat—against the ordinary man in the street, or at least disdainful of him.

Heidegger, in his struggle to follow Aristotle in putting the question of the nature or meaning of being as the central issue for philosophy, saw how little one can say directly and meaningfully about it. That inevitably makes one see that it is first necessary to study the being of human beings themselves, the unique creatures who can pose such issues, since they are already in the world and, certainly as adults, have much knowledge about things, or beings. They know then something about Beingness. So we must start our thinking from man's being in the world—*Dasein*—which already implies a grasp of Beingness, even if a very limited one. There is more than nothing, but what does that imply or mean? Heidegger is saying something about ontology, knowledge and psychology, and most importantly for Boss expressing ideas relevant to psychotherapy. Perhaps especially helpful for Boss was the critique of the scientific bases of medicine and of psychiatry in particular; the scientific set of assumptions within which Boss (1975/1983) had been taught.

After the War, Boss arranged recurrent visits of Heidegger to Zurich. On the first occasion to give an address in the then state of the art, modern, technically well appointed lecture theatre. Subsequently, they conversed in Boss's own house in Zollikon. There Boss invited 50–70 others to come and join them over a period of about 10 years, to interact regularly with Heidegger. Those meetings became well known as the Zollikon Seminars (Heidegger, 1987/2001). They were only terminated when Boss realized that Heidegger was becoming too old and ill to make the trips. The distaste both Boss and Heidegger showed for the technical, and their preference for a more human dwelling reveals the understandable and perhaps the characteristic psychodynamics involved. In many ways Heidegger longed for the traditional, elite, cultured, poetic, philosophical, artistic, aristocratic, humanistic, ethical, high minded, and religious (Heidegger had at one time begun to train to become a Jesuit, and had even converted his protestant wife to Catholicism though he himself left the church to be free to pursue philosophy unhampered by dogma). Heidegger's personal history and loves perhaps explained much of his initial respect for the Nazis. He does seem to have been prone to making the mistake (sometimes regarded as one to which Germans are prone) of "overvaluing culture." In his case, Heidegger focused on the national culture and a language—the wisdom of which he revered and respected, constantly coupling its insights with those of Ancient Greek. The two languages however have little in common, although Heidegger was very familiar with the etymological aspects of both, and he supports many of his arguments by referring to original meanings. He does so in ways that scholars of the classics do not necessarily accept.

Boss wrote that it took him 4 years of the joint seminars at his own home before he really saw how the insights of Heidegger could be useful in clinical psychiatry. However, Boss never seems to disagree with Heidegger in his theoretical statements, and it can be taken that what is offered here as Heidegger's views are also those of Boss. That is, in so far as they are presented correctly. It is very important to stress here that fighting one's way through the obscure writings of Heidegger is akin to trying to walk through a dense thicket; happily Boss wrote more clearly than Heidegger. What we will present here relies heavily on his three most important books, as well as the Zollikon Seminars. The books are the *Analysis of Dreams* (1953/1957), *Psychoanalysis and Daseinsanalysis* (1957/1963), and *The Existential Basis of Medicine and Psychology* (1975/1983).

As Boss himself implied, it is not easy to understand Heidegger and several important writers failed to do so. Ludwig Binswanger, for instance, who was the very influential forerunner of Boss as an existential analyst, accepted, when told so, that he had been mistaken about significant aspects of Heidegger's writings. Sartre also, according to Boss (1957/1963) failed to comprehend Heidegger. That was much more disastrous for the world, for Boss wrote that Sartre caused international ignorance of Heidegger's real achievements, especially in the large parts of the world so intrigued by the French intellectual scene. Sartre gave an idealist and a Platonic account of Heidegger according to Boss (1957/1963, trans., pp. 50 and 51), when much of Heidegger's thought was a rejection of Plato, who he felt added to the failures of European philosophy. Certainly no phenomenology can entertain the view of Plato that appearances obscure realities beyond them which to some extent we already know from our previous existence.

Before pursuing any further the considerable struggle to do better than Sartre and Binswanger, it is appropriate to describe the context of Boss's own training and so to see what it was that he did so want to change, and how Heidegger helped him to do so.

HISTORICAL SETTING AND TRAINING OF MEDARD BOSS

With the above facts in mind, we can consider the actual education of Boss. Also originally studying in Zurich at the Bergholzli, was Ludwig Binswanger

(1881–1966). Binswanger was in fact the first psychiatrist to be profoundly influenced by the philosophical studies of Edmund Husserl's (1859–1938) phenomenology and then by Heidegger, Husserl's most brilliant student. This led to the development of phenomenological psychiatry, existential psychiatry and Daseinsanalysis. Medard Boss, Binswanger's younger Swiss compatriot, knew and was inevitably influenced by Binswanger (there is an admiring account of Binswanger in *The Analysis of Dreams*, 1953/1957). But he himself became, and remained, the more faithful follower and more accurate exponent of Heidegger's output. He was aided by spending many of his holidays together with the Heideggers, especially around the Aegean, and basking in the remains and atmosphere of ancient Greece.

Both of the Swiss psychiatrists had been very privileged. They had been thrown into a creative kernel and internationally dominate period of German-speaking psychiatry. It was around the turn of the twentieth century, that Germany, having previously suffered from repeated French invasions, itself began to become the more powerful country economically. It consequently made the most massive contributions to academic studies, including psychiatry, physics, chemistry, and philosophy. German academic literature was the product of those German by nationality and others speaking and writing in German, including workers in Austria, Switzerland and much of what is currently Hungary, the Czech Republic, Italy and Poland; many Scandinavian and other scientists also published in German. It was for a long time very fashionable for British psychiatrists to quote the German classics but rare to have actually read them as translation was patchy and delayed. As far as psychiatry and many other disciplines were concerned there was already available the French foundations, of nomenclature, taxonomy and theory. As detailed by Jenner and Kendall (1991), the French provided in the nineteenth century the dominant lines of academic study, influenced greatly by the ideologies and outlooks of the French Revolution (1789–1794)). That period is often seen as the culminating victory of the eighteenth century, the "Age of Reason" or as some would have it, the century of destructive cynicism and the watershed for modern psychiatry. The one towards the end of which, in 1781, Immanuel Kant (1724–1804) poignantly wrote *The Critique of Pure Reason* (Kant 1781/1929).

Modern psychiatry is not unreasonably considered to have been initiated by Phillipe Pinel (1745–1826), through his *Traité Médico-philosophique sur l'aliénation mental ou la manie* (1801/1962) so pilloried by Michel Foucault (1926–1984) in *The Birth of the Clinic* (1963/1973). Pinel prospered because he was favored by the revolutionaries. Internationally he displaced the dominance of British "moral treatment," providing his own revised version based on it. His early texts were followed by those of his many illustrious French successors (see Jenner & Kendal, 1991). However, a massively significant discovery in France in 1822 by Antoine Bayle (1799–1858) altered the focus or balance of concern away from Pinel's more humanistic, psychological and political outlook (see Brown, 1994). This produced the fundamental postulates of much German psychiatry, most starkly expressed by Wilhelm Griessinger's (1817–1868) contention that "mental diseases are brain diseases" (see Jaspers, 1913/1963, p 482.) Griesinger had worked in Zurich and maybe his impact was felt later by Boss, though the effect is hard to demonstrate specifically.

Bayle had shown that many of the so-called mad had a clear physical defect of the brain (for him an illness demonstrated at post-mortem by thickening of the meninges, the covering tissues of the brain). These people had dementia paralytica, or general paralysis of the insane (GPI), in fact cerebral syphilis. The triumphs of pathological anatomy and developments in bacteriology, and the apparent victory of reason over religion, and the massively successful technology based on the natural sciences of the times, colored the philosophical thought of the nineteenth century, and certainly German psychiatry. It also terrified many German philosophers well into the twentieth century by the degree to which it might make their work just appendages of the natural sciences. That was a concern reinforced in the twentieth century by the influential Vienna Circle and logical positivism. It is worth noting, however, that the paradoxical attitudes of psychiatrists to the organic verses psychological approaches is well illustrated even by Griesinger, the apparently most hard-headed physiologist. In his 1845 textbook, he has a chapter on the influence of the family!

Boss' dates (1903–1990) made him a product of, or reflection of the twentieth century which his life almost spanned. He has to be seen as reacting to and involved in dominant streams of the Germanic thought of the era. All the time he seems to have been struggling to be a worthy, genuine and authentic human being.

Psychology in particular illustrated the current materialist threats to philosophy, humanistic literature and religion. At the beginning of, and well into the twentieth century, psychology as a discipline was a branch of philosophy, but it too was becoming one of the natural sciences. It was in particular influenced by the medically trained experimentalist and physiologist Wilhelm Wundt (1832–1920; see for example Petersen, 1925). Emil Kraepelin (1856–1926) a leading and historically extraordinarily influential psychiatrist had worked for a while with Wundt and later with many neuro-anatomists, such as Franz Nissl (1860–1919). It is of note that such famous neuro-anatomists as both Nissl and Alois Alzheimer (1864–1915) were actually directors of academic units for clinical psychiatry. German and Austrian (and French)—but not Swiss (nor British) psychiatry—were closely integrated with neurology and controlled by neurologists. So Kraepelin understandably enough assumed that it was reasonable to look for other conditions similar to Bayle's dementia paralytica. As a man of his time he hoped that better studies in anatomy combined with attempts at clinical classification would help in this, at least heuristically. Hence he worked on a classification based on signs and symptoms. It does not seem that he noticed that Bayle's discovery was by post mortem and not clinical examinations. At the end of Kraepelin's life, the reason why what he had attempted had been so difficult became more obvious even to him. There is a limited degree of correlation between physical changes in the cerebrum and the psychological signs and symptoms displayed by the sufferers (see for example Avenarius, 1979). Kraepelin's candidates for comparatively new diseases were in particular manic depressive illness and schizophrenia, which he at the time called dementia praecox. He initially felt that these patients tended to start to be unwell at puberty, as they often do, and didn't ever get better, which in fact many do. His were only comparatively new syndromes though *folie circulair* and *demence precose* among other conditions had already been outlined, although somewhat differently, by French workers.

It was Boss' original chief Eugene Bleuler (1857–1939) who showed that Kraepelin was mistaken to have felt that recovery from dementia praecox (schizophrenia) did not occur. The implication that recovery happened, was that the illness could not simply be the consequence of a permanently altered or abnormal anatomy, in the way that Alzheimer's disease is. Despite this, Kraepelin's clinical outlines (even if not his presumptions) have survived the twentieth century and are persisting into the twenty-first century. The search for physical cerebral concomitants also continues millions of dollars later, now using nuclear magnetic resonance and other very powerful related techniques to find the needle in the haystack which might not (and Boss thought, cannot) be there. Kraepelin's work has been ridiculed but it has not been replaced by anything more adequate. That is so although at the end of Kraepelin's life he, humble scientist that he was, confessed that he had been too categorical, and he showed that he himself had very often failed to distinguish, for example dementia praecox from dementia paralytica. Binswanger and Boss, the ardent anti-reductionists, used Kraepelin's nomenclature while rejecting many, if not all, of his assumptions. Boss (1957/1963) wrote about the features distinguishing between manic depression, schizophrenia, neurosis, hysteria and psychopathic disorders.

Since it is obvious that we all need to use a vocabulary most of which our readers or hearers will understand, communication (and maybe thought) requires a socially produced language created by those Heidegger sometimes scathingly calls "the They." Even Heidegger the great producer of neologisms, sees this very well in relation to the stance of a people, which their background and language has largely shaped. Part of the difficulty in understanding him is the degree to which he alters the meaning of, and creates new words. He defends himself with an etymological excuse few can truly evaluate.

Binswanger and Boss would have been taught the argot of a form of Kraepelinian (almost neurological) psychiatry at medical school, but by being Swiss and not German and by working at the Burghölzli their psychiatry would have been less dominated by pathological anatomists, post mortem studies of the brain and neurology. Boss would have been even less influenced by the powerful organicist schools of Germany than Binswanger, in view of his younger age, and the increasing influence of his chief Eugene Bleuler's (1857–1939) acceptance and respect for Freud. But both Boss and Binswanger used Kraepelin's classification, and both were immersed in the German literature. They would both also have seen patients suffering the dementia paralytica of Bayle, which remained clinically difficult to distinguish from schizophrenia. Penicillin after the second world war almost made cerebral syphilis disappear, as it quickly treats dementia paralytica, while the Swiss psychiatrists advocated psychotherapy for the enigmatic states called schizophrenia, some cases of which for them clearly were not physical disorders of the brain. Boss' contacts within the Burghölzli clinic in Zurich, where he worked and studied, gave him another privileged ringside view of creative developments in the psychiatrically-important German speaking triangle of Zurich (Jung, Bleuler, and Binswanger), Vienna (Freud, Schilder and Wagner Jauregg—who won the Nobel Prize for discovering that malaria treats with some success general paralysis of the insane) and of course Heidelberg, the real intellectual power house of classical academic psychiatry (see e.g., Janzarik, 1971, with organic approaches to psychiatry in the Heidelberg work of Kraepelin, Nissl, Jaspers, Schneider, etc.). Boss did clearly look more to the East, that is, to Freud in Vienna, than to the North, to Heidelberg, for his psychiatric inspiration. Freud, however was not a part of the academic psychiatric circle of Vienna. In fact, that world was critical of his speculative approach.

Boss would have been a junior member of the staff in Zurich when the influence of Freud's ideas was still being discussed, and exposed to the orientation of Kraepelin's approach. Then the comparatively revolutionary outlook from a Viennese upstart was expanding the horizons from those of the more conservative Heidelberg (and the more official academy in Vienna too).

Both Boss and Binswanger were attracted away from the attitudes of the classical psychiatry of Kraepelin by Freud (1856–1939). They were impressed by the possibilities of psychoanalysis. Working and studying as both did at the Burghölzli clinic in Zurich, they were heirs to the excitement of what Carl Gustav Jung (1875–1961) had brought back from his time with Freud in Vienna. Returning to Zurich, it was Jung who had influenced the chief of the clinic, Eugene Bleuler (1857–1939)—who had sent him to Vienna—to take a sympathetic attitude towards Freud and psychoanalysis. Bleuler also felt that Kraepelin's outlook was too pessimistic. The Freudian influence and the disillusion with Kraepelin led Eugene Bleuler to replace the Kraepelinian term dementia praecox with that of the "group of schizophrenias" ("*Die Gruppe der* Schizophrenien," 1911). Throughout the world, that term was fairly quickly to change to schizophrenia in the singular, and hence to a more categorical and homogeneous diagnosis and nomenclature.

Bleuler had tried to produce some tentative rapprochement between Freud and Kraepelin. He actually wrote fairly critically about both perspectives. As a younger man, however, Bleuler had been optimistic about the therapeutic outlook for the person with schizophrenia. He felt they could be helped psychologically in the way Freud believed only the neurotic person can; Freud himself remained doubtful whether psychoanalysis did offer hope for the psychotic individual. Boss and Binswanger agreed with the position of the younger Eugene Bleuler. Yet as an older man Eugene Bleuler reverted to a more organic and fatalistic view. He told the famous ballet dancer Vaslav Nijinsky's (1890–1950) loving wife that little could be done for Vaslav, as he had schizophrenia. But it was the outlook of the younger Eugene Bleuler which was more influential for Boss (and for Bleuler's own son Manfred Bleuler).

PRESUMPTIONS

Writing in this sort of field it is also wise to explain the position from which the author himself is coming. My training was in fact within a similar, though far less prestigious medical perspective—if more biochemical—and then psychiatric training. Philosophically I was, and still am, most powerfully under the influence of Wittgenstein's advice not to ask for the meaning of a word but rather how it is used. Kant (1724–1803) had long since suggested that we must take much to be so, which is not so. This for him was because of the nature of our own minds or psychology. Hence he felt justified in "denying reason to defend faith" from the possible extrapolations from the science of his time. What we seem to know, he suggested, is all about what he called phenomena. They were for him appearances, as distinct from noumena, the things in themselves (*Dingen an Sich)*. Three-dimensional space, linear time, and causality are not discovered by us. They are, Kant thought, the parameters into which we find we have to squeeze our "knowledge," and so much of our experiences. We cannot do otherwise, he thought. So reality, what really is, remains more mysterious than our scientific ideas. This seems the same view as Wittgenstein (1922/1961) expressed by his famous saying, "What we cannot speak about we must pass over in silence." Heidegger and Boss in contrast use the word phenomenon to refer to that which is illuminated by us and so discloses itself in its essence to us in the light we shed upon it. A good example of this is given by Boss (1957/1963, p. 54). He is referring to a dream of a patient whom Jung thought gave grounds for accepting the notion of the collective unconscious. The man who knew no Greek dreamt of a "pneumatic divine." Boss urges us to see this as "the immediate revelation of the divine in the light of a Dasein who—without having any archetype in store—is open enough for the immediate appearance of the divine." There are of course at least three possible explanations of that dream, (1) Coincidence; (2) A collective unconscious, and (3) Boss's view. Do we have a way of distinguishing between these, other than by our general presumptions? Boss would argue that his is without any preconception and merely describes what is presented or disclosed to Dasein. In this case the patient's Dasein.

While modern physics shows how, at least mathematically, we can not only think causally about multidimensional space and time, etc., in everyday life we remain as Kant implied, and so we are constrained and obliged to use the verb *to be* roughly as Kant asserted. While doing so, to a degree, Boss implies we are actually speaking correctly rather than simply inevitably. The radical nature of Boss' position is pellucidly stated as follows: "What were things before there were men?" and "What will become of things when men no longer exist?" Approaches other than the one through an analysis of Dasein are meaningless. For Boss, the human being is the measure and not the measurer of all things. Unless Daseinsanalysis is just one of many ways of thinking and without ontological priority, few attitudes could be more distinct from the current enthusiasms for evolutionary psychiatry and psychology. They attempt to explain what man is while accepting for him uncritically what physical things are. It is central to understanding Boss' outlook to grasp his appeal to us to focus on things and issues themselves. It is Dasein, or the human being, which illuminates realities by the very nature of their *being there*.

Heidegger was concerned to look for "truth" as he conceived it, but not lightly accepting some *façons de parler* which we, as human beings must perforce use for most purposes. One nevertheless feels that towards the end of his life, Heidegger is moving increasingly towards a concept of our truth as dependent on language. He clearly sees what we call the sciences as systematically building on what Kuhn (1962) would later called paradigms. These are preliminary assumptions which Heidegger says orient one to a specific kind of answer to questions which are apparently being posed. The avenues taken produce powerful rather than truthful results. Truth he feels is for philosophy to pursue, but it too must start from everyday human experiences even if colored by culture, history and language. Yet can we be made more than humbled and aware of our limitations, and the backgrounds we accept or are rejecting?

SIGMUND FREUD'S IMPACT ON MEDARD BOSS

Freud (1916–1917) in his *Introductory Lectures on Psychoanalysis* puts well what was attractive to Boss and Binswanger, both such independent thinkers despite their classical medical training. Freud wrote for medical persons:

> Psychoanalysis is not to be blamed for a ... difficulty in your relation to it; I must make you yourselves responsible for it, Ladies and Gentlemen, at least in so far as you have been students of medicine. Your earlier education has given a particular direction to your thinking, which leads far away from psychoanalysis. You have been trained to find an anatomical basis for the functions of the organism and their disorders, to explain them chemically and physically and to view them biologically. But no portion of your interest has been directed to psychical life, in which after all, the achievement of this marvelously complex organism reaches its peak. (Freud, 1916–1917/1953, pp. 19–20.)

What Boss and Binswanger would have liked so much less is Freud's delight, as a would-be scientist, in quoting Heinrich Heine (as he did more than once) on philosophers:

> With his night-caps and the tatters of his dressing-gown he patches up the gaps in the structure of the universe. (Freud, 1933/1964, p. 161.)

Freud overtly had little time for philosophy or to him non-scientific thinking. To emphasize that more specifically, Boss quotes Freud as characterising sciences in precisely the unphilosophical manner which Boss wants to criticize and certainly as applied to human psychology. Here Freud was writing about psychoanalysis:

> Our purpose is not merely to describe and classify phenomena, but to conceive them as brought about by the play of forces in the mind, as expressions of tendencies striving

> towards a goal, which work together or against one another. In this conception the trends we merely *infer* are more prominent than the phenomena we *perceive*. (Quoted by Boss, 1957/1963, p. 30, from Freud, 1916–1917/1943, p. 60. See also Freud, 1916–1917/1953, p. 67.)

Boss comments:

> . . . [T]he Daseinsanalytic science of man and his world asks us for once just to look at the phenomena of our world themselves, as they confront us, and to linger with them sufficiently long to become fully aware of what they tell us directly about their meaning and essence. In other words Daseinsanalytic statements never want to be anything more than "mere," if extremely strict, careful, and subtle descriptions and expositions of the essential aspects and features of the inanimate things, the plants, the animals, human beings, Godhead, of everything earthly and heavenly, just as they disclose themselves immediately in the light of the Daseinsanalyst's awareness. Consequently it would be inappropriate, in principle, to regard Daseinsanalytic statements to be "derived" from factors assumed to lie behind that which is described, or to expect that such statements can be "proved" by reduction to presumptions. (Boss, 1957/1963, p. 30)

Boss rightly refers to the change in the meaning in many statements, of the word, science, in fact from knowledge in general, to that essentially related to the mathmaticisation of everything even beyond chemistry and physics and without regard for the humanities. He is adamant about the absurdity of contemplating the possibility of an explanation of the mental in physico-chemical terms. As that is ludicrous for him, so are the hopes of the reductionists. The latter he claims are less empirical and objective than the Daseinsanalysts. The Daseinsanalysts base their assertions on observations they can call revelations (as they are what beings, things or realities do present or disclose to Dasein, to us, the persons already in the world). To a degree this is a contention that there is no difference between much science and much philosophy, both types of thinkers can be ruled by presumptions. These are avoided by Daseinsanalytic approaches because the latter simply involve reporting what is shown to us. They are not hermeneutic, looking to interpret what is disclosed as really something else. Nor are they constrained by what many call logic, which Ludwig Wittgenstein (1889–1951) suggested may be but a branch of the grammar of the language, and Kant thought was our inevitable way of thinking. Heidegger himself also concluded that as he put it "Language is our mistress." This is among many points at which it is necessary to be aware that while Heidegger's *Being and Time* extols hermeneutics, it is after his "turning" (change of mind, *Kehre*) that interpretation is shunned. Further, that although Boss was attracted by *Sein und Zeit*, it was the later Heidegger whom he knew, after the so-called *Kehre* or turning away from—among other things—hermeneutics.

There can be little doubt that Boss is correct to draw attention to the strangeness of applying the determinism so useful in macroscopic physics and chemistry to the problems of persons, politics, and interpersonal relationships, and particularly to clients in psychotherapy. He is in addition right to chide Freud for his declared belief in psychic determinism, which is consistent with that view and simultaneously makes the mental merely epiphenomenal.

Boss wants us to accept most of the experiences that the patient reports as uncontroversial, and certainly not only as a description of their internal world, the existence of which he denies. For him there can be no separation of their internal and external worlds. The patient is Dasein, being-in-the-world. To think otherwise was, for Heidegger and Boss, Descartes' greatest central blunder. We have no separate *res cogitans* and *res extensa*; these exist only in the conceptualisation of Cartesians—and implicitly in so much European philosophy.

Because of the above Boss very psychologically asserts that what is revealed to us, or Dasein, is presented to us in the light of our own mood. The German word or neologism often used by Heidegger is *Befindlichkeit* from the reflexive verb *sich befinden (Wie befinden Sie sich?* How are you?). Attempts to translate *Befindlichkeit* have varied from English neology, "moodedness," to "state of mind"—see for example the introduction to the translation of *Sein und Zeit* by Macquarrie and Robinson (Heidegger, 1927/1962). The point being emphasized is that we are always in a mental state in relation to the world. We are variously open, or attuned to what is revealed and inevitably we are all to some degree unfortunately closed and prejudiced.

Boss is however not only concerned with the meaning of *is* and *to be*, but also of *I*—the personal—and other pronouns. He writes about sitting in his house in Zurich while thinking about the cathedral of Notre Dame in Paris. He asks where is he, even if his body is in Zurich? (Is there a Cartesianism here in Boss' thought? But see Boss, 1953/1957, pp. 82–83.) Possibly his rumination gives us some understanding of his use of Dasein as "being there." Boss takes the same stance in relation to dreams, and illustrates it in his report of "A strange dream of an urn." He wrote:

> Our dreamer ... did not merely see pictures that were only the reproduction of physical reality. Rather she experienced with all her body and soul a world as completely real as she had ever felt during waking life. How else could she have wondered, on waking up, which was the real world: that of the luncheon table in the dream, of her bedroom? (Boss, 1953/1957, p. 82)

Many might find that account difficult, or unnecessary, if it means more than we do when we say to someone half asleep in a lecture, "Are you on your holidays in Spain?"

One particularly important psychiatrist who was a contemporary, and also sufficiently philosophically sophisticated that he became one of the world's leading existentialist writers, was Karl Jaspers (1883–1969). He had Jewish relatives which gives added weight to the fact that he castigated Heidegger at the end of the war for his Nazi sympathies. More importantly here, Jaspers took what was and is the dominant view of psychiatrists to this day. He searched as most medical persons do for the classical tell-tale signs and symptoms of underlying pathological conditions, or diseases. He also had more time for organic studies especially of schizophrenia, which he said cannot be humanistically understood, so it must therefore be a brain disease.

This approach in psychiatry seems much supported by the successes of psychopharmacology, and some would say modern genetic studies. Jaspers' position is much like that of Wilhelm Dilthey (1833–1911; see Rickman, 1976) in accepting two types of knowledge, mathematical, statistical, and scientific about which one can be fairly certain, and humanistic knowledge which often allows less confidence but is usually humanly more relevant. Again most psychiatrists would agree with his approach to psychiatry, often-labeled phenomenological, as expressed in the seminal work *Die allgemeine Psychopathologie* (*General Psychopathology*, 1913/1963). That remains the most influential book of its type, almost a century later. One needs to be careful though about the word "phenomenological" because, in the case of Jaspers, although it is strictly descriptive and often about experiences and expressions, it seeks the bricks, or recurrent themes, as the building blocks upon which a "scientific" edifice of psychopathology might be based. Boss must have been familiar with the work of Jaspers but he does not often refer to him. That, despite the fact that he did take a somewhat analogous view of Freudian theory. However, Jaspers associated his criticism of Freudian theory with that of Freud's practices which Boss so admired, indeed extolled. (For Boss' affirmation of Freudian therapeutic process, see Boss, 1957/1963, chapter 4, pp. 61–74).

Jaspers had a greater scepticism of the therapeutic impact of psychoanalysis. In addition he was looking for a science of psychopathology and in that he was less devoted to listening to the specific individuals themselves. In fact he hardly practiced clinical psychiatry at all. He was a fairly full time theoretical psychiatrist without patients (see Schmitt, 1979). Jaspers sought the tell-tale sign or symptom that explained the problem or perhaps more accurately revealed the illness. Of course with such an attitude Jaspers focuses on interpretations based on observations. Boss is not interested in such an emphasis because he is content with observation or, as he would put it, disclosure of the reality of the Dasein itself.

Boss is also interested in accepting the great effectiveness of psychopharmacology in reducing "hallucinations" and "delusions." However, he dismisses the classical view of perception. In the latter, electro-magnetic light waves are reflected by objects in a way which affects the retina and then the visual cortex. Boss (1957/1963) asserts that his first lesson to students of Daseinsanalysis is to help them to see the fallacies involved in that classical view of the mechanisms involved in perception and in the explanations they tend to give for seeing. He wrote, for example, descriptively about seeing the yellow house across the yard from the lecture theatre. It is not surprising then that he has little time for two types of knowledge, respectively relevant for two types of mental disorder. One approach is literary and one could acquire it in the Faculty of Arts; the other has to be acquired in the faculties of science and medicine. With this latter know-how one can cure syphilis (including general paralysis of the insane, which did so often present itself as a mental disorder).

Whatever the ontological implication of the above may be for Heidegger, in psychotherapy the psychological impact of agreeing with the patient may be more important and certainly to Boss it seemed therapeutically more effective. Presumably so if the patient has not got syphilis! Agreeing is an agreeable way of behaving. That is a way of acting that is often helpful. It respects the patient's way of being in the world and it helps him to feel himself to be a reasonable person in it. Few with any experience in psychotherapy will have failed to note the impact of being perceived to be of interest, supported and seriously listened to. The patient tends to want you to be on their side in things. Explanatory interruptions explaining everything away can be very disruptive and counterproductive.

Jaspers also felt that psychoanalysis can appear to explain anything by extravagances of the imagination, and with this criticism Boss would have agreed. The problem with Freud for both Boss and Binswanger was, though, more the fact that they saw him as another nineteenth century scientist too influenced by positivism and Darwinism (as was Kraepelin). His practices were exemplary, particularly for Boss, but his axioms were unacceptable. Unlike him they did not want to be conventional scientists, and certainly not psycho-determinists, accepting mechanistic metaphors as truly applicable to man. That approach was unethical and for them totally mistaken.

The ethical is for Boss a feature of reality and Boss would have argued that it reveals, or discloses itself to us. It does not require the concepts of logical deduction or interpretation, nor a superego; it is an aspect of the reality of Dasein—living and being there in this world with others. This, he would contend, is unlikely to be grasped by those so enthusiastic to mathematicise the world, and to practice reductionism in the face of experienced reality. By this position he also side-steps the relevance of David Hume's (1711–1776) "You can't deduce ought from is." He would say you do not need to perform such a deduction. Both disclose themselves to the one who allows them to do so, in the light that Dasein casts upon them. This means that the unperturbed person, with an open mind and without presumptions, will perceive what is so.

A case reported by Boss illustrates his view of the comparative superiority of Daseinsanalysis compared to both the psychoanalysis of the Freudian type and the analytical psychology of Jung. (Both he perceived as reductionist, while granting that he had been helped by Jung to reject the absolute centrality of the reduction of everything to the libidinous.) The report is entitled the "Daseinsanalytically modified treatment of a modern neurosis of dullness and the patient's comments on the modifications" (which is presented as Chapter 20 of Boss, 1957/1963, pp. 273–283). This was about a highly intelligent man, a physician, who suffered from a life long guilt-ridden state. He had often had sensual dreams frequently involving his mother. These were invariably followed by wanton destruction of phallic symbols by a paternal figure. What could be more Oedipal, and suggestive of castration anxieties? Later dreams involved his anatomy teacher who wielded an enormous axe-like instrument with which the anatomist struck at the foundations of a church tower. This seemed to be an effort to destroy the tower and bury the patient. His first period of treatment, which was psychoanalytical, lasted for 4 years. The obvious Freudian types of interpretations seemed intellectually very convincing, but the man remained as unwell as before. That was so until he changed his analyst, then he began to feel a little more relaxed.

The second analyst was a Jungian, and the ambience of the sessions were very cosy, and companionable, and in smoke-filled rooms. They involved, indeed consisted of, intellectual discussions while analyst and analysand sat facing each other. The sessions could last several hours and were almost tutorials on mythologies. His dream of a church tower became a religious symbol. This pleased the patient as he now felt less isolated and more like other human beings with similar racial memories. A time came though when the analyst said he could teach the patient no more. He had now to see himself as an ordinary man neither sick nor deviant. The patient tried. But life was not right for him despite having developed a considerable knowledge of the apparently relevant literature. He became increasingly more obsessional about cleanliness.

Feeling alone, desolate, guilty and hopeless and very empty he reluctantly sought help from another analyst, in fact a Daseinsanalyst. When he presented himself, he said he did so *faut de mieux.* The distressed man said how irrelevant to his situation was reducing everything apparently spiritual to the libidinous, and how equally unhelpful were explanations in terms of an (assumed) archetype in a collective unconscious. (Music to Boss' ears?) The patient commented that Kleist (a well-known German writer) had been consistent in committing suicide after reading Kant's view of the inscrutability of things in themselves.

> Where was there something real and genuine to be found which would make life worth living? (Boss, 1957/1963, p. 275)

The analysand was advised to work without psychology. By this Boss meant that they would set aside the underlying assumptions and interpretations the other analysts had used, and of which the patient was now angrily critical. Nevertheless, he got the patient to lie on the couch in the manner of the first Freudian analyst. (We have seen that Boss strongly favored everything about the couch technique of Freud.) So again the patient had to try without reservation to reveal all that came into his mind, however shameful, or fantastic. Precisely the classical technique of Freud. Dreams began to occur about faeces and toilets and churches. He reported having to climb the bell rope to get away from all the faeces in the church and on the font; and climbing the rope made the bell ring. He suddenly awoke and then heard voices saying he was a "shitter." He also became bedeviled by faecal smells everywhere and he could not eat. He became furious with the analyst for making him so aware of his body, and faeces. He even damaged much in the office, and then rushed home and became catatonic. The analyst took personal charge of his stuporose client day and night. (Such involvement is *not* a classical Freudian response.) On emerging from the stupor the patient threw his arms around the analyst repeatedly saying "Mummy, Mummy, dear, dear" (Boss, 1957/1963, p. 277). In due course, though, the classical return to the couch was possible and the patient thanked the analyst for accepting him in his true bodyliness.

It was suggested in Boss's report that the man's problems had been exacerbated by his admiration for the anatomy teacher who had cynically damaged his faith in God. Then he had failed to be open to what would have been disclosed to him. Hence:

> He had not merely fled from an encounter with a psychological symbol of libidinous or archetypal nature. He had closed himself to the very disclosure of the earthly and the divine itself in their whole immediacy. (Boss, 1957/1963, p. 281).

After his recovery the patient wrote to the third analyst to tell him of the irrelevance of the psychoanalytical interpretations of the libidinous nature of everything he experienced, and the pleasant but useless discussions with his Jungian analyst which were so like school lessons about history and culture. The success he attributed to the approach of the third analyst was found in his acceptance of the divine and filthyness of his experiences and of our real state in life.

> Not a single faecal dream turned up in those two years [of the Jungian analysis], and in the psychotherapeutic discussions we were miles from the small-child state to which I was hurtled back in that confusion I experienced when I was with you. It was though only through it, though, that I could find my rebirth and reach the starting point of my way to maturity. Since everything below remained sealed over, I could not genuinely expand upward either. It is only now that I fully understand how it was that during the last part of the analysis, with you, there was a continuous interweaving of sexual and excremental themes with the religious experiences which gripped me so profoundly. Didn't Nietzsche say somewhere "The higher one will ascend toward heaven, the deeper must one first sink one's roots in the earth, if one is not to be blown over by the first wind that comes along?" (Boss, 1957/1963, p. 282).

We are told that after the termination of the Daseinsanalysis the man married and lived happily with a wife and the four children they had. After reporting the above, Boss again typically and generously questions why, if his ideas are right, it is that some people do recover after undertaking classical psychoanalysis. Much more is, of course, reported, but perhaps central to this chapter is the fact that the analyst did encourage the man to see that our lives are stretched between the earthly and heavenly. I may perhaps comment that once again the analysis has possibly been successful in giving the man a way of seeing things in a language satisfying to him, and his personal need for an acceptable image of the self.

Hopefully the above discloses something of the approaches and debates of the medical and intellectual world regarding psychiatry, into which Binswanger and Boss the highly cultured physicians were, with great privilege, thrown. They admired so much about this world in which they had been reared, had thought, and had lived, yet they needed an escape to a more humane and authentically human world. Boss saw Heidegger in particular as offering a way to preserve his admiration for Freud the medical humanist without selling out to him as a reductionist scientist (which Boss asserted he was not except in his books, and teaching sessions).

There is without doubt still a great problem of doing justice both to the human Dasein and also to our "knowledge" in the natural sciences. Perhaps we would be wisest very humbly to steer clear of any real hope of consistency, despite our great need for it? The deterministic psychoanalyst, or neuro-physiologist has himself to be a human being when he goes home to meals with his family, if he has one. As David Hume put it, be a philosopher if you will but first be a man. Psychotherapists must try to deal with just that, being a person! If at the same time one can feel ontologically confident, it presumably helps as you have the truth at hand, or the support of Heidegger.

It is essential in that regard, and when looking at the writings of Boss, to try to understand Heidegger's concept of truth. This, I think, was particularly helpful to Boss. Heidegger in approaching truth is often really referring to the conditions in which statements and beliefs are possible. He insists that we begin by grasping what are tools (things useful for our purposes) and which we have concern about or need of. But we could not grasp anything about any tool without the world or Being, being disclosed to some extent to us. It is as though a world independent of us and our tools only begins to be conceptually possible, even if mistaken, when that which we had thought was useful turns out not to be so, but yet it can be discussed. So an apparently "objective" science can be created, but it will be one founded nevertheless on Dasein initially seeing things which are tools for purposes at hand. It is our Dasein, being there, which in its light allows a being or beings or things to disclose themselves. The world and us and our temporal structure are presented to us. Our temporality is the movement back from its anticipated future to the past and so to the present.

Boss wrote that

> Things can come forth into openness only in consonance with Dasein's actual attunement or "pitch"... Just as the coloring and the brightness of a physical light determine what can be seen by it, so things are always disclosed in accordance with man's pitch. An individual's pitch at a certain moment determines in advance the choice, brightness, and coloring of his relationships to the world. (Boss, 1957/1963, p. 41).

Boss emphasizes that we do not create the world but we do discern much about beings in it by the openness and the luminosity we are. Boss continues to hammer home the degree of dependence of Dasein, and categories or beings, on each other.

> Human being, as world-disclosure, and the things, which shine forth in the realm of its "there," are so immediately integral that Heidegger can say of the relation between Being-ness and man that this relation supports everything, insofar as it brings forth both the appearance of things and man's Dasein (Boss, 1957/1963, p. 42).

Man experiences long and short hours depending on his concern, he does not live in astronomical or clock time. Central to him is concern.

Despite all of the above, Boss' acceptance of the efficacy of psychoactive drugs is intriguing. For example in his report of "A schizophrenic hallucination in *statu nascendi*" (in Boss, 1957/1963, pp. 219–229), he wrote that the "administration of tranquillizers brought recovery so rapidly that even in the course of his first week the following conversation with his doctor was possible" (p. 219). The man reported having had an hallucination of the sun in his room. He said,

> All the time I had the feeling that my sex organ was connected with the sun in the sky and was being excited by it. If I had lost sight of the sun on the wall, the real sun would have come close to the earth and the earth would have gone up in flames. (Boss, 1957/1963, p. 220; see also commentary in Heidegger, 1987/2001, pp. 151–152).

Boss tells us this man had been a dependent loner still living with his parents but attached to one long term friend who had recently let him down. Boss explained things in terms of Dasein's need for others. All of which might be looked at in different terms. However the success of the pharmacology can be considered in the light of Heidegger's view of sciences as studies on how to control nature, even human nature, without revealing meaning or truth. Psycho-chemistry does show how mental states and even beliefs (delusions) can be chemically altered.

Perhaps such facts can be viewed in the light of what Heidegger (1959/1966, p. 54) wrote about them

> We can use technical devices as they ought to be used, and also let them alone, as something that does not affect our inner and real core... that is let them alone, as things that are nothing absolute but remain dependent on something higher. I would call this comportment towards technology which expresses 'yes' and at the same time 'no', by an old word releasement toward things.

Further

> Thinking does not overcome metaphysics climbing still higher, surmounting it, transcending it somehow or other; thinking overcomes metaphysics by climbing back down into the nearest of the nearest. (Heidegger, 1947/1978, p. 254)

One might be helped to grasp this point by considering in medicine the controlling, even if therapeutic effectiveness of those neuroleptics (anti-psychotic drugs) for schizophrenia, of which Boss seems to have approved. Like aspirin for headaches, arthritis, premenstrual tension, etc. they clearly do work, but less clearly demonstrate the nature of the pathological condition, the truths about the person. Certainly anti psychotic drugs may be as their discoverers suggested *camisoles chemiques* (chemical straitjackets), they control for our or someone else's convenience. In the case of the hallucinated man, though they could have obscured his way of being in the world, at least they made him and it less troublesome to others. Perhaps, without an attitude like that of Boss, many would have had little desire to question the man's being in the world any further. There were of course problems of living still requiring attention, which were not followed up but nor were they denied. The man was not in psychotherapy, he was an ordinary person admitted to a psychiatric ward perhaps lacking the desire or money to pay very much. Most of the patients receiving psychotherapy at length (and who, therefore got reported) were intellectual, educated, verbally skilled and presumably wealthy.

However, both Boss and Heidegger write as though natural science has a philosophical outlook, which many scientists do not. Perhaps they would have been wiser to use the term scientism for the unreasonable degrees of extrapolation seen from some scientific hypotheses to facts, often not yet demonstrated to be true. Heidegger and Boss remained nearer to the presocratic Protagoras holding "that man is the measure of all things." They are clearly correct in seeing Dasein or man as the unique creature aware of his own future end unto death. Dasein also is an awareness of temporality perceiving everything within its frame. Perhaps less ontological and more relevant to Boss's struggles is being able to talk and be with others. This hallucinated and disturbed man had lost the ability to be with the only one other with whom he had felt at home. His being and security in the world had been changed. The breakdown had, he himself added, taught him this need he had and to which he must attend. A point being made is that the genuine person is at home in his genuine living with others; liking, fearing, blaming, and everything else we do and understand.

A Freudian psychoanalyst might have jumped on the sexual references quite freely made by the patient. Boss puts the sexual aspects of the story on one side, with nothing more than some passing reference to the difficulties of saying what is homosexuality. Presumably, he felt that the patient's relation with the one friend was in some way libidinous. He adds that the patient was still only able to reveal some of his closest secrets and those only because of his perception of the kindness of the doctor. Obviously there was more that could have been said had he felt free enough to do so. Despite the understandable curiosity of the doctor it was necessary to respect the persons right to disclose only that which he felt able and willing to disclose. The patient's perception of the analyst's humanity is a recurrent and important *leitmotiv* throughout the writings of Boss. This is a repeated theme possibly explaining his own and many other therapists' successes. Perhaps this is more important than their theories, even Daseinsanalysis, perhaps it really answers Boss's very repeated question about the success of alternative approaches. If that is so, one can see how those who use a method with success accept its apparent effectiveness. Like scientific ideas might it be efficient but not true? Have we really any criteria for such truths better than the simply pragmatic? Daseinsanalysis may well provide the patient with a feeling that the analyst takes the other person's world or Dasein seriously by not telling them that what they say means something else. A good illustration is found in Boss's report of "The patient who taught the author to see and think differently" (Boss, 1957/1963, pp. 5–27). After even bottle feeding the lady as if a baby and staying one whole night by her side, and tolerating her infantile behavior, and even the smearing of her faeces and her desire to act with behavior beyond all social inhibitions imposed on her, the analyst accepted her. Then when she was ready the analysis was continued as formally as possible.

She said her recovery was because she felt that the analyst was concerned for her. The evidence seemed striking that he was. So Boss again generously added that a psychoanalyst might have produced a similar successful outcome. Presumably he too would have had to break some rules of non-involvement until a stricter approach could begin again? The patient had however been furious with the analyst at first when, before she had converted him, he had tried to explain away her hallucinations in terms of the ego defense mechanisms of projection. Hence her continuing in therapy might have been more difficult if she had not won that argument in favor of the Daseinsanalytic approach, which he was then to follow.

Jaspers wrote that he who has not studied his own philosophy is most ruled by it. This was certainly Boss' view of medically trained psychiatrists, but not including Freud. The latter very over optimistically felt that philosophical difficulties presented to some (by the unconscious mind) would disappear with the therapeutic successes of psychoanalysis. Freud believed that pragmatic evidence would make people see that his theories were correct. Unfortunately the value of psychotherapies remains difficult to assess and even when successful, the reason for this remains elusive, as Boss himself stated. However Boss saw Freud, in his practice as not influenced nor ruled by his own presumptions, nor was he the cold indifferent and distant physician he extolled in his writings. In fact he was a warm and caring person. Perhaps the same could be said about Boss and his own allegiance to Heidegger's views?

Heidegger though as has already been stated saw natural sciences as the dogmatic religion of our times with their claim to present the only access to knowledge. The natural sciences have been developed to control and master nature, rather than to understand being. Their search is not for meaning, but for mastery. That is not what a disturbed patient needs, he wants personal support and help to believe in his own personal possibilities, in fact his ability to be, and to realize that we are until death always becoming and with possibilities. This attitude was inspiring to Boss. It must be therapeutic to have the conviction that we all have possibilities, and can aspire to authenticity. Heidegger helped him by declaring a relationship to (medical or psychiatric) science as follows:

> There is no abandonment of science, but on the contrary, it means arriving at a thoughtful knowing relationship to science, and truly thinking through its limitations. (Heidegger, 1987/ 2001, p. 18).

Boss (1975/1983) adds, in fact contrasting Daseinsanalysis with natural science, at least in relation to being human:

> It is the fate of man to be entrenched by and large in the natural scientific tradition of seeing only causal relationships among isolated objects. Thus unusual effort is required of an investigator who would experience the non-objectifiable openness which the Dasein-based orientation shows as the true basis of human nature. The new phenomenological-existential viewpoint cannot be got from reading books, nor can it be achieved through logical deductions from already existing ideas. This viewpoint

> can be reached only through a quantum leap in thinking, feeling and perceiving... (p. 164)

Heidegger had little time for epistemology as a central aspect of philosophy. Our knowledge is given to us. He reported that Kant thought it a scandal of philosophy that it could not give categorical evidence for the existence and knowledge of an external world. For Heidegger the scandal was "such proofs are expected and attempted again and again" (1927/1962, p. 249). Heidegger's concept of Dasein, being there and already in the world, with his insistence that our knowledge and hope of any ontology begins there, is perhaps more helpful to the psychotherapist who needs a philosophy than the questions of unreal scepticism of so many other philosophers. With Daseinsanalysis Boss does not depart from the world the patient does know. How ridiculous can one be to ask about where one is, when in fact already among and using the things one doubts that exist. The sentences, though, raise the question of the meaning of the verbs to be and exist, and so of being.

Heidegger does argue that as far as we are concerned the antique tables and chairs and hammers or tools of our everydayness are that from which our knowledge begins and on which it rests. Much of virtue in his writing depends on his insistence that we should pay particular attention to our initial grasping of the nature of things. Then we see the virtue of his stress on the unhiddenness of reality. Truth does not depend on the correspondence of statements to things in themselves, rather language itself is in charge of us and is especially seen to be so in his later writings.

Boss repeatedly insists that Freud did hold moralistic positions which are meaningless in terms of his own theoretical deterministic assumptions. Clearly Boss and Heidegger saw man's irreducible moral dimension which is revealed to him in his openness which allows an illumination of his being. Guignon (1993) explores the moral issues in psychotherapy well. Even at an empirical level, self-respect and mental health seem to demand some ethical criteria. Most would accept that claim, without needing to objectify it—or seeing it, with Heidegger and Boss, as what reality discloses itself to the openness of Dasein. The aesthetic by which we must live does not need to be ontological truth. Perhaps unfortunately it can be a reflection of our herd instincts as higher animals. It would still be important therefore in psychotherapy. It would help one to see how foolish it is to go against our own nature.

Despite Heidegger's great love of poetry especially that of Holderlin and Rilke, Boss was anxious to scotch any idea that Daseinsanalysis is just some metaphorical approach to reality. And Boss takes reality as a whole: he is of course focused on human existence and behavior but he does not distinguish the world of the psyche from another outside world. That leads one to reflect on the view of Rorty (1993). He suggested, influenced by Wittgenstein's (1953) concept of language games, that we should thank Heidegger for giving us yet another language game to play. This is one which Heidegger gave to us, but which he (and Boss) thought Being had given to him! Perhaps?

SUMMARY

What then is the heritage left by Medard Boss? There is the great support for Freud's release of psychiatry from the simplistic medical axioms of much of the nineteenth and twentieth centuries. There was also admiration for Freud's technique of the couch and the essentially silent therapist, offering very infrequent remarks to help the relaxed clients to see for themselves that they have potential. Beyond this, Boss showed that there is no need, within psychotherapy, for Freud's lingering faith in the ontology of nineteenth century physics, determinism and the natural sciences. Nor was Freud correct in believing that the disorders he regarded as "narcissistic neuroses" (manic depressive states, and schizophrenias) are not amenable to psychological treatments.

Was it necessary though, to completely follow Heidegger's audacious concepts of a fundamental ontology? Could not as much have been achieved using Husserl's study of a pure intentionality, bracketing what may be so outside of our direct experiences? That would have put enough of contemporary physical science and its struggle to be ontological on one side. To do that is clearly necessary in psychotherapy. However perhaps a religion also helps the psychotherapist?

Boss had high ideals for therapists, who

> must not intervene even in the interest of the therapist's own God to seek to guide the partner's life. ... The author himself has more than once been forced to admit that he was not ready to open himself to an analysand sufficiently to be able to live up to the Daseinsanalytic demands made upon him. Whenever he found himself in this situation, he referred the patient to another analyst who was capable of meeting the demands of the patient. Invariably the patient improved in an amazingly short time with the new therapist. (Boss, 1957/1963, p. 260)

Boss' work and life has inspired many others. For most, his therapeutic approach was very humane, concerned, and impressive, the underlying philosophy was also interesting, yet for some, problematic, and not so necessary.

REFERENCES

Avenarius, R. (1979). Emil Kraepelin seine Personlichkeit und Konzeption. In *Psychopathologie als Grundlagenwissenschaft.* Stuttgart: Enke.

Bleuler, E. (1911). *Dementia praecox oder Gruppe der Schizophrenien.* Leipzig: Deuticke.

Boss, M. (1957). *The analysis of dreams* (A. J. Pomerans, Trans.). London: Rider. (Original work published 1953)

Boss, M. (1963). *Psychoanalysis and Daseinsanalysis* (L. B. Lefebre, Trans.). London: Basic Books. (Original work published 1957)

Boss, M. (1983). *Existential basis of medicine and psychology* (S. Conway & A. Cleaves, Trans.). New York: Jason Aronson. (Original work published 1975)

Brown, E. M. (1994). French psychiatry's initial reception of Bayle's discovery of general paresis of the insane. *Bulletin of the History of Medicine, 68,* 235–253.

Foucault, M. (1973). *The birth of the clinic: An archaeology of medical perception* (A. Sheridan, Trans.). New York: Vintage Books. (Original work published 1963)

Freud, S. (1943). *A general introduction to psycho-analysis* (J. Riviere, Trans.). New York: Liveright. (Original work published 1916–1917) [The translation cited by Boss.]
Freud, S. (1953). *Introductory lectures on psycho-analysis* (J. Strachey, Trans.). London: Hogarth Press. (Original work published 1916–1917)
Freud, S. (1964). *New introductory lectures on psycho-analysis.* (J. Strachey, Trans.). London: Hogarth Press. (Original work published 1933)
Guignon, C. (1993). Authenticity, moral values and psychotherapy. In G. C. Guignon (Ed.), *The Cambridge companion to Heidegger* (pp. 215–239). Cambridge: Cambridge University Press.
Heidegger, M. (1962). *Being and time.* (J. Macquarrie & E. Robinson, Trans.). Oxford: Blackwell. (Original work published 1927)
Heidegger, M. (1966). *Discourse on thinking* (J. M. Anderson & E. H. Freund, Trans.). New York: Harper. (Original work published 1959)
Heidegger, M. (1978). Letter on humanism (F. A. Capuzzi, Trans.). In D. F. Krell (Ed.), *Basic writings: Martin Heidegger* (pp. 213–265).London: Routledge. (Original work published 1947)
Heidegger, M. (2001). *Zolliken seminars* (M. Boss, Ed., F. Mayr & R. Askay, Trans.). Evanston: Northwestern University Press. (Original work published 1987)
Janzarik, W. (1971). *100 Jahren Heidelberg Psychiatrie.* Stuttgart: Enke.
Jaspers, K. (1963). *General psychopathology* (J. Hoenig & M. Hamilton, Trans.). Manchester: Manchester University Press. (Original work published 1913)
Jenner, F. A., & Kendall, T. J. G. (1991). The French revolution and the origins of modern psychiatry. In D. Williams (Ed.), *1789: The long and the short of it* (pp. 99–120). Sheffield: Sheffield Academic Press.
Kant, I. (1929). *Critique of pure reason* (N. Kemp Smith, Trans.), London: Macmillan. (Original work published 1781)
Kraepelin, E. (1903). *Psychiatrie ein Lehrbuch.* Leipzig: Johann Ambrosius Barth.
Kuhn, S. (1962). *The structure of scientific revolutions.* Chicago: Chicago University Press.
Petersen, P. (1925). *Wilhelm Wundt und seine Zeit.* Stuttgart: Fr. Fromanns Verlag.
Pinel, P. (1962). *A treatise on insanity* (D. D. Davis, Trans.). New York: Hafner. (Original work published 1801)
Rickman H. P. (1976). *Dilthey: Selected writings.* (Especially chapter "Ideas about Descriptive and Analytical Psychology") Cambridge: Cambridge University Press.
Rorty, R. (1993). Wittgenstein, Heidegger and the reification of language. In G. C. Guignon (Ed.), *The Cambridge companion to Heidegger* (pp. 337–357). Cambridge: Cambridge University Press.
Schmitt, W. (1979). Karl Jaspers und die Methodenfrage in der Psychiatrie. In W. Janzarik (Ed.), *Psychopathologie als Grundlagenwissenschaft.* Stuttgart: Enke.
Wittgenstein, L. (1953). *Philosophical investigations* (G. E. M. Anscombe, Trans.). Oxford: Blackwell.
Wittgenstein, L. (1961). *Tractatus Logico-Philosophicus.* (D. F. Pears & B. F. McGuiness, Trans.). London: Routledge and Kegan Paul. (Original work published 1922)

CHAPTER 9

CONTEMPORARY EXISTENTIALIST TENDENCIES IN PSYCHOLOGY

STUART HANSCOMB

HOW EXISTENTIALISM AND PSYCHOLOGY MEET

Modern psychology's relation to existentialism takes at least three discernible forms. Firstly, theory and practice (particularly in psychotherapy, but also in other areas of psychology) that is directly traceable to philosophers and other writers who are termed "existentialist."[1] Secondly, theory and practice that is not traceable in this way, but which raises questions, uses concepts or interprets findings in ways that are markedly existential. (Here, it is often the case that the researchers in question are unaware of the connection.) Thirdly, apparently non-existentialist psychological ideas—say, naturalistic, or psychodynamic insights—that are (coincidentally or not) found in the works of existentialist philosophers.

In the first category—direct influence—it is useful to distinguish between philosophers who were not in any strict sense psychologists, and philosophers whose subject matter is sometimes explicitly psychological. Of the former kind those most commonly cited include Kierkegaard, Nietzsche, Heidegger and Merleau-Ponty; and of the latter kind, Sartre, Buber and Tillich. One thinker who has written both purely philosophical and purely psychological works is Karl Jaspers (Jaspers, 1913/1963), but he is an oddity in this respect in that, although initiating a phenomenological methodology, was scientistic.[2] Another category of influences are thinkers who are not themselves usually labeled as existentialist, but who share various themes and ideas with the usual canon. This includes psychoanalysts like Rank (for his emphasis on the will and creativity) and Lacan (for, among other things, his placing the ego on a more intersubjective footing), and philosophers such as Wittgenstein, Ricoeur and Foucault.

Some landmarks in the development of existential psychology and psychiatry are, in Europe, Binswanger's Heidegger-influenced *Dream and Experience* (1930) and Boss' Heidegger influenced (and approved) *Psychoanalysis and Daseinsanalysis* Boss (1957/1963, see Jenner, this volume); and in America May, Angel, and Ellenberger (1958). With articles by Binswanger and Minkowsky, along with contributions by the editors, this latter book signaled the beginning of the existential psychology movement in the States. In the early 1960s May and Adrian van Kaam started the *Review of Existential Psychology and Psychiatry*. Notable among the contributors and editors over the years are Paul Tillich, Gabriel Marcel, Viktor Frankl, Leslie Farber, R.D. Laing and Thomas Szasz.[3]

The most famous name in contemporary existentially-oriented psychology is Irvin Yalom (who studied under May). His *Existential Psychotherapy* (Yalom, 1980) remains, to my knowledge, the only truly comprehensive textbook on the subject and, like several of the philosophers who have influenced him, he has also written fiction.[4] In Britain the current most influential institution is the Society for Existential Analysis that was founded by Emmy van Deurzen in London in 1988.[5]

In the second set of existential tendencies in contemporary psychology—theories and findings that are noticeably existential but which do not draw directly or at all from existential philosophy—there is a broadening of relevant subject matter. It is certainly the case that existentialist themes emerge in non-existential forms of psychotherapy, and within many of the other sub-disciplines such themes and ideas are evident. Examples include models of lifespan development (such as Gould's and Levinson's (Gould, 1978; Levinson, 1978)); work within cognitive and social psychology on "cognitive dissonance," the "self-serving bias" and "depressive realism," and theorizing about the nature of emotions.[6]

Related to this point is a general willingness among existential philosophers and psychologist to draw on a wide range of material. Yalom, for instance, says that "the existential paradigm has a broad sweep: it gathers and harvests the insights of many philosophers, artists and therapists about the painful and redemptive consequences of confrontation with ultimate concerns." (1980, p. 486)[7] In other words, existentialism might be a relatively new and quite specific movement within philosophy and psychology, but existential concerns have always been around, and thus there are few limits on where and when they can show themselves.

The final form of the connection between psychology and existentialism—seemingly non-existential psychology in existential texts—will not receive as much attention over the course of this chapter, but there are three points I want to make here regarding this relation.

The first is that it is certainly true that a great deal of psychologizing goes on in existential philosophy. There are several reasons for this: firstly, questions about the self and about ethics are basic to its subject matter, and under most philosophical banners it is relevant to both of these. Secondly, distinctive to its approach are not only subjectivity and an enticement towards individuality, but an analysis of why these aspects of existence tend to be avoided or forgotten. Existentialism is, in other words, interested in how and why fear and anxiety motivate us and these investigations will inevitably stray into psychological territory of less immediate relevance. Thirdly, and perhaps most importantly, it deals with *concrete experience*. Put these together and you get questions like "Why, in terms of the lived experience of the individual, are questions like 'how should I live?' not fully addressed, and what does this tell us about human being?" Existential philosophers are looking for answers in terms of deep, universal, or necessary truths about the peculiarity of human existence, but the raw material they must work with in order to reach these answers is the concrete experience of the individual. Only through detailed consideration of human behavior and the conscious and unconscious motivations that lie behind it, can they hope to find clues as to the ways in which our basic metaphysical condition affects us. The ostensibly non-existential psychological insights and observations found in the writings of Kierkegaard, Nietzsche, Heidegger and so on are, then, likely to be instances of behavior that can readily, even if not directly, illuminate our existential condition.

The second point I want to make concerns a particular dichotomy in existential thinking—one that will be addressed later in the chapter—between models of the human condition that are developmentally sensitive, and those that are not In brief, the former implies that existential awareness is something we have to grow towards and that the further we are from this awareness the less it can be said to be an influence upon us; while the latter implies that existential concerns can always be meaningfully construed as determinants of affect and behavior, no matter what our age or experience. For now the important point is that although the former might not be any more psychological than the latter, it tends to be psychological in ways that would be familiar to more than the existentially oriented. This, of course, could be a fluke—a result of categorial boundary-smudging in a time philosophers were psychologizing more liberally—but my feeling is that there is a close and predictable relation between these models and the different approaches to psychology they involve.

The last point concerns existential psychology's status as a paradigm (as opposed, say, to a sub discipline). As a paradigm it presents many psychological phenomena in a light distinct from competing paradigms (e.g., evolutionary, psychoanalytic, cognitive). In this respect some of the psychological subject matter under consideration will have been explained in other ways, and, for the reader, might be *better* explained in other ways. (Sartre's redescriptions of unconscious phenomena in his chapters on "bad faith" and "existential psychoanalysis" are good examples.[8])[9]

Its scope is effectively spelt-out by Adrian van Kaam (1990). He stresses that existentialism is relevant to those levels of functioning that differentiate humans from "objects"—essentially intentionality. This means that in some areas of psychology—e.g., aspects of learning theory, animal psychology, and physiological psychology—existentialism is not going to be treading on too many toes, but that in others—personality, emotions, social, and abnormal psychology and their developmental correlates—it is likely to be making claims which seek to usurp clearly competing paradigms (e.g., psychoanalysis), and keep others in their place (e.g., evolutionary theory). Along these latter lines van Kaam says "a central task of a comprehensive existential psychology is the discovery of existential constructs which can integrate the contributions of the various differential psychologies" (1990, p. 23) With respect to the issue of developmental sensitivity, this integration, as we shall see, sometimes just involves a mapping out of the psychological territory in light of the human being's basic and necessary intentional awareness of itself *as* mortal, free, alone and absurd; and sometimes it involves seeing "differential" psychologies as representative of levels or stages on the way to this recognition.

In the remainder of the chapter I intend to address, in one form or another, all three of the ways in which I see existentialism and psychology meeting, though priority will, naturally, go to the first. To begin with I shall tackle existentialism's basic themes and describe and analyze how the self is conceptualized in light of them; and then I shall consider the methodological dictates that arise from this conceptualization.

"Fundamental Concerns" and "Existential Dimensions"

A list of typical themes and terms from existential philosophy would include alienation, freedom, death, despair, absurdity, anxiety, bad faith, and authenticity. In broad terms the existential psychologist is interested in the anxiety that is generated by basic features of human existence such as freedom and responsibility, death, contingency, and aloneness; the ways in which we attempt to deny this anxiety, and forms of life in which it is authentically integrated into a self that is vibrant, self-aware, committed and autonomous.

As mentioned, the most systematic account of the relation these existential themes have to psychology is Irvin Yalom's *Existential Psychotherapy*. In this he outlines and analyses what he calls our "four ultimate concerns"—death, freedom (responsibility and willing), isolation, and meaninglessness.[10] Of these, and like Otto Rank, Paul Tillich, Norman O' Brown and Ernest Becker,[11] he sees death as the most basic, but other writers, though pretty much agreeing on what the basic concerns are, order them differently. Martin Buber and Eric Fromm for instance see isolation from others as our basic source of anxiety; for Camus and Frankl it is meaninglessness, and for Sartre and Farber it's freedom. These four are deeply interconnected which means that in order to analyze them you can start anywhere and sooner or later illuminate them all; but it can further be argued that they are incoherent as genuine existential concerns if treated in isolation from one another. What this then suggests is that there is an encompassing whole that they represent the parts of, but what is this whole? The short answer is conscious human existence, and there might be no further reduction possible beyond this point. Less brief, but perhaps no less mysterious, is to say with Heidegger and others that human being is essentially "uncanny"—a form of being that condemns those fitting the description to being dissatisfied metaphysicians; to always be, in Thomas Nagel's words, formulating more questions than answers can be provided to (Nagel, 1986, Ch. 11). I shall return to this matter at various points in this essay.

A less systematic psychological approach to existentialism is that of Emmy van Deurzen-Smith (For example, see van Deurzen-Smith, 1997, Ch. 18). The differences between her and Yalom (and other, particularly American, existential psychologists) are significant and I shall look more closely at these when I consider methodology, but certain similarities in the way she subdivides human existence provide further confirmation of common roots in the existential tradition. Specifically, she does not talk about "fundamental concerns" and their attendant anxieties, but more neutrally of "existential dimensions." These she terms the "physical," the "social," the "psychological" and the "spiritual." They refer to the individual's embodiment, to her relation to others, her relation to herself and to the meaning of life as a whole. Yalom's ultimate concerns clearly map onto these, but as we shall see, the work they do for him differs in some important respects to the work these dimensions do for van Deurzen-Smith.

In my summary of how existential philosophy has influenced psychological notions of the self I intend, ostensibly at least, to map-out the self with reference to these anxieties and the dimensions they represent. I do not, however, want to claim that there is a *primary* concern among those that Yalom lists and begin there; instead I want to follow the lead of a recent commentator on existential philosophy and say that the experience that makes best sense of all these concerns is uncanniness, or a sense of being "not at home." At certain points my approach joins up with van Deurzen-Smith and other writers in underlining the deeply paradoxical nature of human existence (the reason, she claims, that her approach is genuinely existential and that Yalom's is not).

The commentator in question is David E. Cooper, who in his "reconstruction" of existentialism claims that the principle concern that links philosophers tagged "existential" is *alienation*. It is not the sort of alienation that has come about only as a result of "recent historical circumstances" (World Wars, technology, dissolute middle classes, etc.); and it is not the sort that will be overcome by the inexorable unfolding of "Spirit" through history (Hegel), nor by changes in economic circumstances (Marx). Rather, it is an alienation brought about by inappropriate dualisms (e.g., mind and body, fact and value) that infect our world-view and which, in the beginning at least, need to be philosophized away (Cooper, 1990, pp. 31–36).

I agree that there is philosophical work to do here, but the manifestation of alienation of particular interest to psychologists is slightly different. We are, I believe, often alienated from ourselves, not only because of faulty world-views, but because the world the existentialists will have us accept—one where "the world... is indelibly human; and humans are indelibly worldly" (Cooper, 1990, p. 81)—is itself one where we struggle to feel at home in quite the way we might like to.

The process of acceptance is as much a personal one that takes commitment and courage as it is an intellectual, reflective one (which is partly why existentialism has caught the imagination of artists and psychologists as well as philosophers).

If we accept that some form of separation and "uncanniness" or "not-at-homeness"[12] is an inevitable part of the human condition we can see why some form of anxiety is also part of that condition. Indeed, one of the best accounts of homelessness and our responses to it is found in Kierkegaard's *The Concept of Anxiety*. His "psychologically orienting deliberation" uses religious categories like faith and sin, but in many respects the features of pre-religious existence he describes capture universal existential structures and experiences. In what Kierkegaard calls the "qualitative leap," the forming individual recognizes his distance from the infinite through a profound subjective realization of his limitations as an embodied person. Deeply ignorant of their self and the world, and yet seeing that to have a self of any substance they must take responsibility for their finite (particularly sexual) nature and actions, the individual experiences a vertiginous insight that is at once a glimpse of the infinite and a sense of helplessness before it. Homelessness, in short, is for Kierkegaard the inevitable result of our being a "synthesis of the finite and the infinite,"[13] and, shorn of religious implications, something similar is being described by many or all existentialists. As well as Heidegger's uncanniness, the human situation and the possibility of an authentic response to it throws up terms like "ambiguity" (Kierkegaard, de Beauvoir), dizziness (Kierkegaard, Camus), meta-stability (Sartre), insecurity (Tillich) and irony (Rorty). It is summarized by Merleau-Ponty in the following way:

> Our birth . . . is the basis both of our activity and individuality, and our passivity or generality—that inner weakness which prevents us from ever achieving the density of the absolute individual. We are not in some incomprehensible way an activity joined to a passivity . . . but wholly active and wholly passive, because we are the upsurge of time. (1945/1979, p. 428)

It is for this reason then that anxiety is the favored choice for many existentialists for describing our primordial response to our condition. It is not the sort of anxiety that in itself can be alleviated, and nor should we seek to repress it. It is an anxiety that demands a response from us—to live authentically—and on-going attempts to deny its significance create forms of life that are in some sense stunted—Sartre's "bad faith," Tillich's "unrealistic" self-affirmation, certitude and perfection (Tillich, 1952/1962, esp. Ch. 3), or Heidegger's "they-self."

At this point in the analysis psychology gets a clear foothold. Once a need, and thus a motivation, is in sight, human behavior can be investigated in terms of responses to this need. The four concerns can be linked in this way. Isolation from others makes us existentially anxious because relationships of all kinds can, on the one hand, create an illusion of canniness which satisfies our basic metaphysical craving; and on the other they can serve, in the moment, to obliterate that craving. Anxiety associated with individual freedom follows as nothing is more telling of our separateness from others than the recognition that only we can take responsibility for our life and that in a critical sense we are the author of that life. In turn, avoidance of freedom's anxiety means our lives are not fully lived, and an absence of engagement or commitment is certainly a cause of lives felt to be meaningless and, as Norman O. Brown has observed (and many have agreed[14]), "the horror of death is the horror of dying with unlived lives in our bodies." (Yalom, 1980, p. 151)

Facing the source of anxiety, most would agree, does not disempower it, but it can alter the experiencing of it. After Heidegger, van Kaam says that for "the healthy person who is able to face and accept his contingency in openness, this anxiety is pervaded by a peaceful, humble acceptance of this aspect of human reality." (van Kaam, 1961, p. 211) This must be seen as a fundamental aim of existential therapy.

Other People

Unavoidably we live in a social world and unavoidably we are, in part, socially defined, but what is the existential relevance of other people? Two broad concerns arise from this—concerns tackled comprehensively by Sartre, Buber and Marcel. Firstly our "locus of control" with regard to freedom of many kinds including our self-definition; and secondly the temptation to use relations with others as a way of avoiding our freedom and uncanniness. This second I shall deal with under the next heading—"freedom, guilt, and death."

To resist being defined by others is healthy up to a point, but there is also a point where its inevitability must be accepted. In portraying human relations in *Being and Nothingness* as essentially involving conflict, Sartre is saying more than that we are always vulnerable to the restricting labels, moral condemnation and malicious gossip of others (subject matter familiar to social psychologists); he is saying that even with the best intentions relationships will become a battlefield, At bottom, what we find disturbing is the very truth that other people's views on us (and even their gaze on us) is something we cannot avoid or ignore.

Why don't we like it? One issue is mastery. The consciousness of another means that "the world has a kind of drain hole in the middle of its being" (1966, p. 343)—in fact endless impenetrable holes that make our environment unpredictable and dangerous. But this is perhaps nothing more than an interpersonal problem—something that might create paranoia as well as moral and practical problems and the requirement for sensitivity in communication, theories of mind, empathy and a balance between suspicion and generosity.

Sartre's suggestion however is that we also find this state of affairs ontologically disturbing. Confronting other minds (whether it's someone's glance, a suspicious lover, or even the categorization implied by one's social identity) is a reminder of the fact that the world is intersubjective at its core: the meanings we live among are *human* meanings and that is all they could ever be. But why should this create anxiety? There are at least two reasons. The first is a pseudo-existential one—we are reminded of our smallness among all the many billions who have lived and who will ever live. (This is not in itself a cause of anxiety or despair, rather, as Nagel, points out, it is the clash between this perspective, and an everyday one in which we take our pursuits very seriously, that we can find unsettling. (Nagel, 1979, Ch. 2)) The second is that an intersubjective world is not a *necessary* world with all the assured sense and purpose that that would seemingly bring with it. Perhaps both these sources of anxiety amount to the early Sartre's summation of the human condition as a "useless passion"—a hopeless "fundamental project" to be both creator and creation; like God, both free and necessary, and essentially impossible.

In *Being and Nothingness* (and other works) Sartre tells a naturalistic, often psychological, story alongside the philosophical one. The chapters on "being-for-others" in particular read more like case studies in pathological relationships than illustrations of necessary interpersonal dynamics that inevitably arise from our "situated" existence. This causes some quite serious interpretative problems, but the psychologist can be less concerned by these and let herself be informed about modes of behavior that might signify attempts to avoid existential anxiety. The situation Sartre presents is one in which to quell our uncanniness we either seek to dominate the other or allow them to subsume us. Both strategies seek a kind of immutability—the interpreting other is silenced or we lose ourselves in them—and both, even where the other's strategy is compatible, will necessarily fail.[15] These poles are of course extremes, and they demonstrate the distance we can go in flight from anxiety; but Sartre's message is also that as an essentially intersubjective self we can only ever be sliding towards one or the other. In the space between (say) two people, there can be no merging of consciousnesses and thus no neutral field where the real, essential me and the real essential you can freely graze. In fact the whole idea of real, essential personalities becomes non-sensical—something Walter Mischel's situationist critique of trait theory underlines. (See, for example, Mischel, 1968) The authentic individual recognizes this necessity and her resultant "meta-stability" and non-oppressive interaction with others Sartre investigates in a later work. (Sartre, 1983/1992).

Yalom lists a number of ways, some similar to Sartre, in which we seek to obliterate the anxiety caused by the rupture between ourselves and others (these include "existing in the eyes of others," "fusing" with others and sexual compulsivity (Yalom, 1980, pp. 378—391, 1991/1989, Chapters 1 and 2); and Sartre has famously influenced Laing's views on interpersonal relations. *Self and Others*, for example, includes accounts of the type of destructive pre-emptive spirals that are created when each party attempts to avoid the pain of objectification by first objectifying the other (see also *Knots*); and in *The Divided Self* Laing describes analogous fears on the part of the schizophrenic in terms of "engulfment" and "petrification" (Laing, 1968, Ch. 3, 1969, Ch. 3).

Much of the material written and cited by existential psychologists concerning authentic relations with others shares many, often commonsensical, views with differently oriented theorists. (Fromm, 1957, and humanistic psychologists like Maslow (1987, Ch. 12) and Rogers (1961/1967, Ch. 18) are mentioned often enough.) By far the most influential existential philosopher in this regard though is Martin Buber. The distinction he draws between "I-thou" and "I-it" relationships expresses the difference between being truly present for another and engaging with them instrumentally encountering them, deliberately or otherwise, from an individual's particular point of view and on that individual's terms. They are assimilated into pre-formed schemata and agendas that disallow the emergence of the "delicate," naturally independent, and ultimately positive "between." Indeed, what Buber seeks is more than the Kantian moral imperative of treating people as ends in themselves: he says,

> When two men converse together, the psychological is certainly an important part of the situation, as each listens and each prepares to speak. Yet this is only the hidden accompaniment to the conversation itself ... whose meaning is to be found neither in one of the two partners nor in both together, but only in the dialogue itself, in the 'between' in which they live together. (Wheway, 1999, p. 123)

Buber stresses the primacy of intersubjectivity ("in the beginning is the relation"(Buber, 1922/1970, p. 69)) and his ideas on therapeutic practice have had the greatest impact on psychology. (See especially Buber, 1947) "Dialogical therapy" (Friedman, 1989; Wheway, 1999) emerged directly from his work, and existential therapists, including Yalom and van Deurzen Smith, often cite his influence.

Just how dialogue heals I shall say more about under "methodology," but for now there is a crucial existential twist to add to the authentic intersubjective picture. Something that is less of an issue for Buber, but critical for many existentialists, is the matter of the tension created between the need to sustain individuality and the development of the potential of our (necessary) relations with others. Important analyses and remarks on this have come from Kierkegaard (notably 1843/1985, 1842/1987), Heidegger (1926/1990, especially Division Two, pt. IV), de Unamuno (1912/1954), Marcel (1949, 1951), Sartre (1983/1992) and de Beauvoir (1948/1994); and Cooper (1990) provides an incisive summary of how the problem can, in part, be solved. The essence of this position is that since our measure of ourselves is inextricably linked to how others view us, a requirement of our maintaining our authenticity is that we help maintain authenticity in others:

> an authentic understanding of myself as freedom ... requires me to view others as possessed of this same kind of existence. Unless I so view them, I cannot expect them to view me in that manner. Only if I regard and treat others—or better, regard them *through* treating them—as loci of existential freedom, will I receive back an image of myself as just such a locus. (1990, p. 189)

Different authors have different thoughts on how this is to be achieved and on how far one should go in one's attempts to achieve it. Kierkegaard, for instance, has said that "the most resigned a human being can be is to acknowledge the given independence in every human being, and to the best of one's ability do everything in order to truly help someone retain it" (1846/1992, p. 260); and de Unamuno that "true charity is a kind of invasion ... it is to awaken ... uneasiness and torment of the spirit." (1954, p. 282) More aggressive forms of therapy like that developed by Ellis ("Rational Emotive Therapy") might fit de Unamuno's bill, but gentler contrivances can be found in therapeutic literature. Richard Hycner, for instance, distinguishes between I-thou "moments" and I-thou "processes"—the latter being "purposive intervention which, conducted respectfully, support the conditions for I-thou moments to occur." (Wheway, 1999, p. 114) It may not be quite this that Hycner has in mind, but in his play *Emergency*, where a psychiatrist in need of help but refusing to seek it is therapized indirectly by another psychiatrist pretending to be his patient, Helmuth Kaiser demonstrates the sort of imaginative intervention existential analysts would not automatically dismiss.(Yalom, 1980, p. 253)[16]

As said, in non-professional relationships existentialists' views on authentic love are not radically different from other paradigms.[17] The person truly in love will do all they can to abet the other to be all they can be—they are thus absolutely *for* them but not absolutely *with* them except in *necessarily* impermanent I-thou moments. Their reality and value is often not in question, but the true existential significance of such occurrences lies in what surrounds them. Through bravely confronting our uncanniness and not seeking refuge in, or power from, the other, we are better able to create and discern genuine I-thou moments and I-thou relationships. These are good *in themselves*, but of equal (or greater) value is individual freedom. Fortunately these are mutually supporting goods, and since the former can only arise indirectly the individual should be committed to their own authenticity and, sometimes through I-it style contrivances that do not directly infringe on the others' self-determination, the authenticity of the other.

FREEDOM, GUILT, DEATH

Avoidance of freedom causes guilt, and guilt causes further avoidance of freedom. Why are we guilty, according to the existentialist? Because we have not taken a grip of our lives and fashioned them creatively in terms of the meanings that encircle the "narrative center of gravity"[18] that is our particular self. Using Virginia Woolf's metaphor, we have too often looked the other way as the drops of self continually form and fall (Woolf, 1931/1977). That is one source of guilt, but not the only one and not, I would say, the most clearly existential one.

This other I will come to in the section headed "meaning and absurdity," and for now I will consider this first source of guilt. Existential psychology is concerned with "potential," but it needs to be careful identifying what kind of potential this is. Clearly it is not a given nature or set of characteristics or traits that we need to "actualize." When Maslow says we strive (or should strive) "to be all we can be" (1954/1987, p. 22) he is not being quite the essentialist the slogan suggests, but that he would use it at all is nevertheless an indicator of a divergence in the existential and humanistic paradigms.

For the existentialist potential generally means the potential to see ourselves as free in the relevant senses, as opposed to, say, a mere synthesis of biological determinants and environmental conditioning. Similarly, just as Nietzsche attacked what he saw as oppressive Christian morality, so the modern existential psychologist rails against the individual conceived of as the hapless victim of immature (e.g., Oedipal) guilt. The question for the psychologist though is how the individual is supposed to achieve freedom. On the face of it responsibility and authentic relations with others are hard won, but why is it so hard? and to what extent is it itself an act of will?

The traditional morality closest to the existential orientation is virtue theory. This is because certain virtues like courage and commitment seem necessary in order to achieve one's "existential potential." Thus facing our anxiety and living authentically require, indirectly, the development of character strength.

Relevant here is Leslie Farber's "two realms of the will." Decisions rationally worked through and conscious efforts of will comprise the second realm, but choice, of such importance to the existentialist, is not always conscious. This for Sartre would be a dissatisfying state of affairs, a culture suited to bad faith, but for Farber (and tacitly or otherwise for many existential philosophers) it is of vital importance. His point is that meaningful changes of lifestyle and personality often needs cultivating in advance. Small, authentically motivated, alterations of habit combined with moments of self-reflection and rational thought will work subintentionally towards that change. The individual may well be unaware of just how much is going on until confronted by a circumstance that tests them. Then, without any need for a "dead heave of the will,"[19] they act in such a way that reveals, retrospectively, how different they have become. "I can will knowledge, but not wisdom" says Farber, "self-assertion . . . but not courage." (Yalom, 1980, p. 299) A process like this is existential and authentic because the force behind it has been the individual's determination to change themselves and/or their world-view; the existentialist can accept without contradiction the oil tanker nature of character and habit so long as she believes change, freely desired, is possible (Farber, 1966).[20]

That there are polarities within existential psychology with regard to the exact relation character strength has to existential authenticity is revealing in more ways than one. A simple way to express the nature of these poles is in terms of the question: Is authenticity to be equated with character strength, or is character strength a necessary prerequisite for authenticity? The former implies that character development is hindered only, or primarily by existential anxiety; the latter that it can be hindered by (potentially many) layers of other factors before it has to encounter existential concerns. Yalom steers towards the former: death anxiety in his account of our condition playing a not dissimilar role to the libido in Freud's. From an early age, he claims, children are in denial of death (1980, Ch. 3), and like Tillich (1952/1962),[21] he sees most of our adult neuroses as having a similar origin. Symptoms exhibited work for the patient either by confirming their "specialness" or the existence of an "ultimate rescuer." (Yalom, 1980, Ch. 4)

Two criticisms that can be leveled at Yalom are these: first, even if death anxiety is in some sense basic, it is not necessarily always an *existential* anxiety. By his own admission death anxiety in adults is mitigated by a meaningful life,[22] but until adolescence at least, children presumably do not conceive of their lives in quite this way. If we also assume issues of isolation and freedom are also not existentially flavored for them, then can the death anxiety that does exist really be of the order he wants it to be? This criticism applies to adult fears of death as well. Death, I accept, provokes an anxiety like no other, but I am not convinced that, by itself, it is a truly existential anxiety. Thoughts of our own death, I suggest, provoke a primordial terror; one that if we try to rationalize it creates something we might call anxiety, but not for the reasons the other "ultimate concerns" do (i.e., our inherent ontological instability), but because, as Tillich would say, of the *ungraspable* nature of our non-existence. "It isn't *natural* to die," says a character in Sartre's story *The Wall*; "man has invented death" says Yeats (*Death* (1974, p. 142)).

Death's existential significance should perhaps be sought elsewhere. I've mentioned its relation to one sense of a life lacking meaning (fear of it is all the more acute when we feel our lives have not been lived to the full); but it also provides a potent reminder that life is not the sort of thing that can have the kind of overall, externally ordained purpose and plan we might yearn for. Death does not come at a time when we have done all we can do and completed our task (or perhaps our selves), not because we have limited control over when it comes, or even because it *must* come, but because, as Rorty (1989, p. 42) says, "there's nothing to complete." A life is not like a jigsaw puzzle or a mathematical equation. Relatedly, what is death if not the ultimate separation? On the one hand it's a fundamental part of our constitution and yet something that we have little control over; and on the other it's something we must do alone.

If I am right and it is this context that makes death an existential concern, then Yalom and other theorists need to show that this context is present even from an early *adult* age. This is the second criticism and it takes us in the direction of the opposite pole to Yalom's reductionism. It is not clear that our anxieties become anything like existential until we have matured and explored ourselves and our world. They might, firstly, be present but vague; and secondly the existence of the features of existence that give rise to existential concerns may well be metaphysically woven to the types of situations that do cause us concern, but this is far from claiming that we are always, consciously or unconsciously, existentially oriented. In the first instance we might, for example, intellectually grasp life's absurdity after reading *The Myth of Sisyphus*, but our maturity and life circumstances might be such that we are far from ripe for it to have "heat" (to paraphrase William James). In the second, it is a truism to say that if it were not for our freedom and isolation from others there would not be relationship crises, but this does not mean that the relevant people are, or even could be, aware of the role they play.

My point is that there are other, more or less self-contained realms of concern that are in an important sense incommensurable with full-blown existential concerns. I will, though, immediately qualify this in the following way: these could (and I believe they do) form something like a hierarchy (maybe, traditional criticisms not withstanding, akin to Maslow's "hierarchy of needs") with existential awareness somewhere near the top.[23] When and if the hierarchy is climbed (and I would also say there is something like an existential gravitational drag that forces us to do so) nothing on the lower rungs will look the same again unless understanding is repressed. Crucially then, this polarity is not saying that existential concerns and our methods of denying them are key shapers of personality and pathology from the ground up; they are key shapers only when life has taught us enough for the whites of their eyes to be in view. Once seen we then have the power to re-assess much of what has gone before (hence, in part, the value of existential therapy for adults), but until seen, until life's uncanniness seeps far enough in to color our predominant world-views, it makes no sense for it to serve as the fundamental psychological paradigm.

If I am reading her correctly, van Deurzen Smith is closer to this "developmental" pole (there is more on this aspect of her approach below), and the naturalistic aspects of Nietzsche and, in particular, Kierkegaard seem to fit this picture (notably Kierkegaard's "stages on life's way"—the aesthetic, the ethical and the religious[24]). It is here also that we can locate a cross-over with models of lifespan development that have been formulated since Erikson in the 1950s (for example, Erikson, 1950) and particularly over the last thirty years (for example, Levinson, 1978). The "static" model on the other hand, at least with regard to adulthood, is closer to Heidegger and Sartre. (With the overtly psychological Sartre, the absence of developmental considerations is, I think, conspicuous.)

Authenticity and Integrity

Since authenticity is on the one the hand existential-knowledge and acceptance and on the other self-knowledge and acceptance (as conditioned by existential knowledge), then the criticisms and detailed descriptions of modes of self-deception found in existential literature are to be expected.[25] Almost invariably the message is to face ourselves and our existence bravely and resolutely and not shirk any of its truths or demands.

Avoidance of our "being," like avoidance of internal conflicts in psychoanalytic theory, elicits metaphors like rigidity and inflexibility.[26] Nietzsche's "four errors," Laing's schizoid retreat, the "neatly drawn map" of Hesse's "pseudo-world," William James' instinct to "absolutize," Tillich's "unrealistic certitude" and Kundera's "kitsch" are examples.[27] Many (notably James and Heidegger) accept that some kind of pure and continuous existential awareness is not possible, but all seem to agree that for one's boundaries to be drawn up in a particular way for too long a time is a sign of inauthenticity. (In the next section I will describe in more detail a particular type of flexibility the existentialist demands of us.)

Where a more static psychology is implied (as with Sartre), avoidance of this primarily involves an act of will like Farber's second "realm"; but where a developmental model is implied the picture is both more complex and more compatible with a "psychological realism." Alexander Nehamas (Nehamas, 1985), commentating on Nietzsche, describes a version of authentic integration that would, in spirit at least, be accepted by many personality and clinical theorists:

> A self is just a set of coherently connected episodes, and an admirable self, as Nietzsche insists again and again, consists of a large number of powerful and conflicting tendencies that are controlled and harmonized. Coherence, of course, can also be produced by weakness, mediocrity, and one-dimensionality. But style... involves controlled multiplicity and resolved conflict. (1985, p. 188)[28]

Inextricable from the Nietzschean ideal is a courageous existential awareness—the power to affirm "all that is questionable and terrible in existence" (1968, p. 39)—not made apparent in this quote. Despite unresolved methodological and theoretical questions regarding the locating of this person in his general psychology, many of the important features of this ideal are found in Maslow's descriptions of "self-actualizers." For instance:

> Our healthy individuals find it possible to accept themselves and their own nature without chagrin or complaint ... They can accept their own human nature with all its discrepancies from the ideal image without feeling real concern [but] it would convey the wrong impression to say that they are self-satisfied. (Taylor & Brown, 1988, p. 196)

And one commentator on Maslow has said,

> Perhaps the most universal characteristic of these ... people is their unusual ability to perceive other people correctly and efficiently, to see reality *as it is* rather than *as they wish it to be*. They have a better perception of reality and more comfortable relations with it ... [they are] able to tolerate ambiguity and uncertainty more easily than others. (Hjelle & Ziegler, 1981, p. 388)

An attempt to discredit this aspect of mental health has emerged from research into learned helplessness and attributional styles during the 1970s and 1980s. Some of the findings culminated in a paper by Taylor and Brown (1988) that made the claim that realistic perceptions are detrimental rather than basic to healthy mental functioning. Examining the vast amount of research into our "self-serving bias," including evidence that the attributional styles of depressives are significantly more realistic than those of nondepressives (in terms of their assessment of how much control they have over events, degree of optimism about the future, and how popular, talented, etc. they are), they concluded that "the capacity to develop and maintain positive illusions may be thought of as a valuable human resource to be nurtured and promoted." (p. 193)[29]

There are many criticisms—both conceptual and concerning external validity—that can be made of this paper and the research that preceded it, but I will mention just a couple here.[30] Firstly, it is not the case that *all* non-depressives are overly-optimistic, so what are we to make of those who have slipped through the statistical net? Relatedly, the claim here is not that realism *causes* depression, but that realistic people tend to be depressed and optimistic people tend not to be. Even though some research does suggest an absence of positive illusions is a significant predictor of depression, some of the statistics Taylor and Brown base their claim around can be accounted for by reversing this causal process, and equally, one assumes this can be done with the optimistic group. Factor this in and it could leave us with a far weaker pattern of results and greater reason for suggesting that there are plenty of people whose realism has not caused mental health problems.

My second point is that not enough account is taken of the distinction between short-term defensive devices and more detached, long-term self-perceptions. This is particularly relevant to the types of competitive laboratory tests often employed to study perceptions of control; tests that are often only paying attention to subjects' immediate responses and then under somewhat peculiar conditions.[31]

Both these points underline something seemingly crucial to the whole business of authenticity: it may well be that plenty of us employ a self-serving bias to keep ourselves buoyant, but then this is not surprising if we accept that it is extremely tough to achieve authenticity. There may be a great deal to come to terms with before we can afford to maintain a consistently accurate appraisal of our averageness. More to the point, perhaps only by struggling towards an existential perspective can we amass enough of the right kind of objectivity to forgive ourselves our failings and be less concerned about the relative abilities and judgments of others. The preceding implies a developmental model (and in this respect the degree to which students are relied upon as subjects in these studies is bothersome[32]), but it also implies that Sartrean acts of good faith (in one sense of the term) and Heideggerian moments of anxiety and resoluteness are themselves vital parts of the developmental process. What it *also* implies is that there are going to be times on route to authenticity where things seem confused, odd and dangerous, and perhaps it is no wonder that the "journey" metaphor is not an uncommon one in existential philosophy, literature and psychology.[33]

My point is that existential philosophers have rarely made any bones about how difficult authenticity is, and if we accept that a courageous attitude towards the "ultimate concerns" is inextricably linked to a courageous attitude towards our limitations as social, embodied selves and to the chanciness of life, then it should come as no surprise that the studies in question have not uncovered many non-depressed realists.

MEANING AND ABSURDITY

Sometimes when existential psychologists talk about meaning they are referring to how fulfilling a particular person's life is—in an important sense, how *happy* they are. Yalom, for instance, lists eight "secular activities that provide human beings with a sense of life purpose" (1980, p. 431), including altruism, dedication to a cause, creativity, and self-transcendence, but agrees with Viktor Frankl (and Aristotle) that happiness itself is not something that can be pursued directly.

The critical idea is *engagement*, and all the time we are estranged from our day to day activities this is absent. Engagement's relation to meaning is two-fold: one is the energy with which we pursue something. Focus, concentration and effort take us out of ourselves (or at least solipsistic versions of being "in" ourselves) and facilitate the type of interaction with the world familiar to Zen Buddhists[34] and analogous to Buber's ideal relation to others. Frankl's logotherapist ("meaning-centered" psychotherapist) is "in the first place ... concerned with the potential meaning inherent and dormant in all the single situations one has to face throughout his or her life." (Frankl, 1959/1984, p. 168) The existential guilt discussed above arises through the individual not engaging with, and thus not making the most of and learning from circumstances they find themselves in, including tragic ones. The process is self-sustaining: the less we engage the less we learn about what we do and do not value and thus the more we remain in ill-fitting situations (jobs, relationships). This creates more guilt which through our working at avoiding it makes us increasingly neurotic or "hardened"; the hardening means we authentically engage even less with our circumstance and so on.

The second relation is, then, the nature of the activity itself; it must either be something worth doing for its own sake or with such an end (maybe distant in time) clearly in mind. Yalom's list amounts to aspects of life that we tend to find valuable in this way—often pursuits in clear contrast to the typical Western wealth-pursuing grind—but there are endless activities that have the potential to be meaningful in themselves.[35] Meaning, in this respect, depends on the person, and in particular it might depend on where that person is developmentally. Broad ideas like "creativity" and "commitment" are qualities individuals can find in many different places and often the most a therapist can do is make suggestions arrived at through getting to know the person in ways they don't quite know (or have forgotten) themselves. Quite often certain stereotypically meaningful or noble pursuits just don't move us, and if they do they might only be central to our lives for a limited amount of time. The point is that (1) as individuals the passion and commitment expressed in our engagements is crucial; (2) different activities are not all equally worth pursuing (but in part this can only be discovered through committing to them); (3) the criteria for what is worthwhile are strongly dependent on the individual in question.

Frankl speaks of three "avenues" to meaning: work (or "deeds"), other people, and self-transcendence in terms of one's attitude to one's fate. The first two are tied in with the type of meaning just discussed, but the third is somewhat different and I would say more distinctly existential. As well as the limitations of our interactions with others, Sartre and others make it clear that we can neither be identified with our social roles nor with what we create (however necessary and important these are). Frankl says,

> we cannot understand the whole film without having first understood each of its components ... Isn't it the same with life? Doesn't the final meaning of life, too, reveal itself, if at all, only at its end, on the verge of death? And doesn't this final meaning, too, depend on whether or not the potential meaning of each single situation has been actualized to the best of the respective individual's knowledge and belief? (1984, p. 168)

If we have been engaged with our lives' bigger patterns, meanings are likely to emerge as we get older, and death can in that respect be happy or guilt free. But the existentialist is also saying that there cannot be a "final meaning"—that it's artificial when books and films offer one. In this respect life is "meaningless," but only in a way in which we cannot truly make sense of what "meaningful" could mean except through necessarily vague references to art and god.

Just what is disturbing about life, our lives "as a whole," is expressed to some extent by Camus in *The Myth of Sisyphus*, but I believe is best captured by philosopher Thomas Nagel (1979, Ch. 2, 1986, Ch. 11). He conceptualizes any rational creature in terms of a polarity of subjective and objective points of view. The subjective is importantly defined in terms of its values, passions and engagements; the objective is impersonal and can view these pursuits from varying degrees of detachment. Epistemologically we can of course never entirely leave our subjectivity, but we can go far enough to take fairly objective stock of our emotional life and particular commitments. It is from this point of view that they seem relatively small and arbitrary but, because emotional categories do not apply here, this does not induce a sense of despair. What it does reveal is the oddity of this duality, and the "absurdity" of life refers not to its absence of metaphysical meaning, but to the fact that being able to take this point of view does not alter our level of commitment to our arbitrary lives.[36]

Problems arise, existentially speaking, because the same faculty that allows us a detached point of view is manifest in the practical reason we use in our day to day lives. Through engagement we learn about specific situations with their own internal sense and value structures, but as we have seen, key to the existential importance of engagement is that it provides us with leverage to accept or reject commitments in light of who we take ourselves to be. Authenticity thus requires reasoned detachment, but there is nothing to stop this reasoned detachment climbing (maybe rocketing) towards a profoundly objective point of view whereby life choices can no longer be grounded in self-justifying projects. The purest meaning of existential anxiety is I believe this—that the more committed we are to our lives the more exposed we are to a point of view that is in conflict with the nature of this commitment.

In existential philosophy and psychology there are various ways in which this pure sense of anxiety is manifest, and often (and often misleadingly) this is referred to as "guilt." One is the reluctance we can have to committing ourselves to someone or something because of an awareness that other, potentially as or more rewarding, alternatives have to be foregone. In Heidegger's words, "freedom ... *is* only in the choice of one possibility ... in tolerating one's not having chosen the others and not being able to choose them." (1926/1990, p. 331) Prevarication and avoidance of conscious choice-making is often a temptation, but the dictates of freedom and authenticity only offer us a stark choice between forgoing autonomy and dignity or committing and taking responsibility even when ignorant and unsure.

Impulsiveness is another way of attempting to defuse this type of "guilt"—something that is itself facilitated by what Yalom calls "sequential ambivalence"—experiencing one desire and then the other and acting upon (or internally opting for) the one that is currently in the foreground. The case study of Mabel who, though in love with her husband, falls for another man, is instructive (and a not dissimilar hypothetical example is used by Flanagan in his essay *Self Confidences* (1996)). She is caught in a cycle of sequential ambivalence that (seeing the clear advantages of one, then the other), though it alleviates anxiety in the short-term, does not provide the opportunity to make the kind of choice she needs to make. Yalom says,

> The therapist's task is to help the impulsive patient turn sequential ambivalence into simultaneous ambivalence. The experiencing of conflicting wishes sequentially is a method of defending oneself from anxiety. When one fully experiences conflicting wishes *simultaneously*, one must face the responsibility of choosing one and relinquishing the other. (1980, p. 313)

He admits in the same passage that "simultaneous ambivalence results in a state of extreme discomfort," but is optimistic about the benefits of facing it. "Creative" solutions might emerge that in a sense solve the problem,[37] but not always, and in many cases we have to always live with thoughts of what "could have been." How to deal with this? Studies in "post-decisional" cognitive dissonance demonstrate our tendency to denigrate unchosen alternatives once we have acted, and this perhaps has to be seen as psychologically pragmatic.[38] However, what should not be ignored from an existential point of view is this necessarily tragic feature of life.

A further manifestation is a situation that was alluded to earlier. Until we commit we do not know, but through committing we are eating away at our own limited time, and weaving our way, often deeply, into the lives of others. The blade is twisted by the requirement that we then take responsibility for these choices, even though in many cases we do not and cannot know where they will take us (and other people[39]). This is particularly true of the adolescent and the young adult; people viewed by society as more or less responsible, but with limited experience. Not only will they not tend to be good at predicting the consequences of their actions, they will not be alive to the full *meaning* of those actions. (For example, what does a legal career mean to the 21-year-old graduate?; what does love mean to the 17-year-old engaged to be married?)[40] We can will authentically and with only good intentions, but there cannot be a guarantee that the results will be good and when they are not, the locus of responsibility still lies with us.

This problem is intractable and is a version of one of the oldest stories told in the existential canon—the "fall" (see Kierkegaard, 1944/1980). Adam's sting was that he could only discover the meaning of right and wrong by *doing* wrong, and so by the time its full force becomes known to him he has, of necessity, already sinned.[41] Existential interpretations of the story of Oedipus salvage it from Freud's narrow interpretation and refloat it in the realm of "the individual's struggles with his fate, in self-knowledge and self-consciousness." (May in Hoeller, 1990, p. 170)

Emmy van Deurzen-Smith cites Heidegger's challenge to Oedipus (in *Introduction to Metaphysics*) of "achieving authenticity in the light of the tragic givens of his life", for, she says, "what is overwhelmingly tragic, is not that he desires his mother and wants to kill his father, but that he has committed all these ignominies whilst believing himself to be doing all the right things." (1997, pp. 182–183) The point is that though doing wrong is an unavoidable part of our condition, we must nevertheless take up these wrongs as our own, take responsibility for them, or else forfeit having a self at all. The self, good and bad, is in many ways a matter of luck, but this luck is the only material we have to work with in order to *fashion* a self.

For the most part what we are not permitted is to choose which aspects of our past go into the mix (only the pouring and shaping of that mix), but youth and experience *are* to some extent mitigating circumstances, and it is the therapist's job to sift authentic guilt from excessive and damaging self blame. The balance required is similar to Yalom's "simultaneous ambivalence"—what we find toughest to endure is not the darkness of death nor the absence of God, but the ambiguities inherent in our freedom and isolation, the necessary lack of stability. Our hardest job, existentially speaking, is to endure meta-stability, be "half-sure and whole hearted,"[42] a "confident-unconfident." (Flanagan, 1996) When it comes to taking responsibility for our past we need not only to avoid the temptations of flippant dissociation and ungenerous self-loathing, but recognize also the nature of existence such that there can be no firm resolutions to such problems.

In general the existentialists' emphasis on the link between emotion and the self is understandable for several reasons. First, the disruptive nature of emotions makes one's cares—what one loves and fears—apparent and vivid. As the "patterns of salience" (de Sousa, in Rorty, 1980) in our lives gain heat, so events gain texture and contour and it is harder to avoid the concrete significance of the decisions we make. Second, in an important sense we are assailed by our emotions; they *happen to us* and can take us by surprise. In this way they are perhaps the most pronounced reminder of our engaged, 'thrown', matter of fact relationship with what is and is not significant for us. Similarly, and thirdly, the physiological and expressive components serve as reminders of our inescapable embodiment; and fourthly, any reflective investigation of the complexities of an emotional episode will reveal the *particularity* of our history, hopes and desires. The particular emotions and moods that existentialists pay attention to—anxiety, guilt, love—are those which capture the *objective* nature of human being in a way that it becomes of deep relevance and concern *for this particular human being*.[43]

"Appraisal" oriented psychologists like Magda Arnold, Nico Frijda and Keith Oatley,[44] as well as philosophers like Nussbaum (1990, 2002), Solomon (1976/1993) and de Sousa (1980) explain the significance of emotions in similar ways. Even neurologically oriented research (for example, Antonio Damasio (1994, 1999/2000)) seems to have features in common with Heidegger's notion of "attunement" (*Befindlichkeit*).[45]

METHODOLOGY

How, according to existential philosophy should people be studied? Immediately it needs to be said that even though some existential psychologists carry out and refer to one or other form of empirical research, all recognize that the important thing is to *change* people and that even though change can only be engendered by insight, that insight has, in a critical sense, to be the *subject's own*. The basic tenets of existentialism are there in the philosophy (with psychologists of course diverging in their responses to, and appropriation of the different biases of these philosophers) and the primary methodological questions concern when and how to *communicate* these insights.

Vital to understanding the existentialist's approach to understanding and handling people is the notion of *subjectivity*. Subjectivity has two important but distinct meanings in existentialism. The first is what Kierkegaard has called "truth as subjectivity"[46] and concerns the *depth and quality* of an individual's grasp of an objective truth. This is what Rollo May is driving at when he describes the purpose of therapy being that "the patient experience his own existence as real." (van Deurzen-Smith, 1997, p. 157)

The second is a more radical reminder of a basic premise of existential phenomenology—that all constructs and values are *human* constructs and values. Scientific truths are not the last call on "reality," these truths are *ideas*, and ideas are found solely in (or between) minds. The correspondence of these ideas with "reality" is not what matters, because even if we accept the special meaning of scientific truths as opposed to, say, moral truths, the situation we find ourselves with is one where, existentially phenomenologically speaking, *all* ideas are on an equal footing. And what this means is that the way the various features of existence are understood, composed, ordered and prioritized by a given culture or a given individual (and indeed which features figure in the first place) is dependent upon nothing other than the history of that culture or individual. Within the boundaries of our imagination the possibilities are limitless, and we can see that there are forms of life we can't, from our historical and cultural positioning, even begin to imagine.

These two meanings of subjectivity give rise, as you would imagine, to two distinct features of existential therapy. (Although these of course are not independent of one another: firstly, among the aspects of existence the individual has to subjectively appropriate is the idea that they are in a crucial sense nothing more than their interpretations and so they cannot expect therapists and theorists to provide answers; and secondly, although they are "free" in this sense there are nonetheless certain things which are unnegotiable such as this freedom itself and the requirement of subjective appropriation.)

First I shall deal with truth as subjectivity. Telling people about the ultimate concerns and the existential dimensions will not, for the most part, make much difference. For this reason empirical research is unhelpful (although it might be helpful for other reasons): what is instead needed are techniques that encourage the individual to understand for themselves the significance of their existential condition.

The same applies to existential philosophy. To an extent what these philosophers are saying is not difficult, and nor is it contentious except perhaps with regard to *how much* existential concerns matter to an individual life. When on occasion Kierkegaard or Nietzsche seem to be over-stating the obvious they are not trying to inform the "reader as philosopher" of anything, but rather motivate the reader as "an existing indiv." towards greater authenticity. This is why something that can be formally stated quite quickly and simply is told and re-told in the form of fables, anecdotes, diary extracts, literary analysis and so on. This, as in the case of Kierkegaard, might be over the course of one volume (e.g., *Fear and Trembling's* many re-workings of the story of Abraham and Isaac), or it might involve themes spelt-out analytically in one book and demonstrated via stories in others (e.g., Camus' linking of his theoretical works like *The Myth of Sisyphus* with novels and plays like *The Outsider* and *Caligula*; and features of Sartre's philosophy such as bad faith, commitment and desire in *Being and Nothingness* given greater life and relevance in *Nausea*, *Dirty Hands*, *The Age of Reason,* etc.). The point is that in order to have the desired effect, the concerns existentialists consider to be basic to human life must be communicated *indirectly*.

In terms of therapeutic approach this necessitates a departure from psychoanalysis. That reflective understanding of the problem is not itself enough—that everyone must *practice* new ways of seeing and living in order to make deep and lasting changes—is not a unique insight of existential psychology. But there is a profound difference in the manner they seek to achieve this, in particular in the way that existential therapists play down the importance of transference and seek to engender a more "real" relationship with the patient in the therapeutic setting. "It is" says Yalom "the relationship that heals" and he cites what is now quite well-known evidence that results are produced in therapy when this relationship is warm and empathetic as well as non-judgmental and accurate (1980, p. 401). Existential extensions of this idea include the therapist serving as a model for the patient in terms of their own (authentic) approach to themselves and to the therapeutic relation, and this can (or should?) involve them being quite candid about their failings. Such candidness prevents the patient feeling that authenticity requires super-human capacities, and also makes them less likely to overstep the boundaries of the relationship and try to off-load inappropriate responsibility onto the therapist.[47]

Existential therapists are generally suspicious of therapeutic "techniques," believing these to impede the spontaneity of the encounter. Few though would go as far as Thomas Szasz when he says that "the relationship... must be as natural... and unrehearsed as is the relationship between other persons who respect... each other" (in Hoeller, 1990, pp. 16–17); and few would use as many techniques as Yalom.[48] His emphasis on the virtues of group therapy as a chance for patients to, among other things, view their problematic interpersonal tendencies in vivo is less of a technique and squares well with many existential aims and assumptions.

Where the departure is more radical is in terms of the *types* of insight existentialists are after and just how these can be appropriated. The ultimate concerns, as far as they are existential, are abstract and difficult to maintain a grip on within the messiness of everyday life, and yet they need to become anything but that in order to make any difference. Worse still, even in its abstract form *the* ultimate concern of alienation seems virtually impossible to ride for any length of time. To at once belong and not belong in the world;[49] to be both a subjective point of view and an objective point of view, to "have to deal with human reality as a being which is what it is not and which is not what it is" (1966, p. 100) is an affront to practical reason, and even if we have the courage to endure the vertigo, direct engagement either with pure reflection or with practical reason in everyday life means we slip from the "dizzying crest" (Camus, 1941/1975, p. 50).

This inherent openness is translated by the existential psychologist into a respect for the individual's point of view. How this cashes out is not simple though and different theorist and therapists have treated it differently. Few are likely to regard the individual's private world as unnegotiable, and in practice no therapist can regard it this way (it is after all in some sense "not working" for the patient); but a dogmatic and scientistic account of conscious and unconscious processes such as Freud's has to be rejected. The therapist must, in a deep sense, treat each case on its own terms, and always be prepared for surprises.[50]

Sartre (1943/1966, p. 716), writing on "existential psychoanalysis" says,

> What we are demanding then . . . is a *veritable* irreducible: that is, an irreducible of which the irreducibility would be self-evident, which would not be presented as the postulate of the psychologist.

The patient's account of her self and her relation to others might be confused and incoherent, but it is for the therapist to clean it up and "reframe" it rather than impose explanatory systems that bear no resemblance to the patient's world-view. As such a key ability of an existential analyst is flexibility, itself the product not of understanding a complex theory of the mind but of "having a rich and varied life" (van Deurzen-Smith, 1997, p. 220).

> The existential psychotherapist has as [her] primary task to recognize together with the client the specific tensions that are at work in the client's life. This requires a process of careful scrutiny and description of the client's experience and a gradually growing familiarity with the client's particular situation and stance in the world. To understand the worldview and the states of mind that this generates is to grapple with the way the client makes meaning, which involves a coming to know clients' values and beliefs. The particular circumstances of the client's life are recognized, as is their wider context. The psychotherapeutic process of existential therapy is then to elicit, clarify and put into perspective all the current issues and contradictions that are problematic (van Deurzen-Smith, 1999, p. 232).

The "perspective" into which these issues are put are the basic existential concerns or dimensions, and the work these do for the patient or "client" depends on the theorist in question. No paradigm or approach is assumption-free and we have seen what existential psychologists tend to treat as given. There are disagreements though, and in particular the one which is exemplified by the divergent approaches of Yalom and van Deurzen-Smith. Yalom, as we have seen, is more reductionist and this perhaps explains the more traditional aspects in his approach to therapy. van Deurzen-Smith is more "top down" and this means giving priority to the "client's" particular and irreducible "world-view."

What perhaps every existential therapist is after is for this perspective to provide, not a "cure" for past wrong-doings and misfortunes or for life's ambiguity, but a sense of dignity and meaning that account for these and prevent them from overwhelming. This is philosophical in part, but it is also highly personal and in an important sense *creative*. The artist, for Rank, "spews" out the world in his work whilst the neurotic "chokes on his introversions" (Becker, 1973, p. 184)[51], but gentler means are also at our disposal so that without resorting to denial we can lift ourselves up, take some control over our lives, and look upon them ironically and affectionately.

NOTES

1. For a brief history of the origins of this term see Cooper (1990, Ch. 1).
2. Note, for example, his comments in *General Psychopathology*: "if [the psychotherapist]... turns to the efforts of modern existential philosophy and uses these ideas as a means of *acquiring psychopathological knowledge*, making them an actual element of psychopathology itself, he is making a scientific error." (1963, p. 776).
3. For more detailed histories of existential movements in psychology, see, for example, Keith Hoeller's Introduction in Hoeller (1990); and Emmy van Deurzen-Smith (1997, Part II, Section 1).
4. See *When Nietzsche Wept* (1992), Lying *on the Couch* (1996). *Love's Executioner* (1989) and *Momma and the Meaning of Life* (1999) are both collections of "tales of psychotherapy." They are not strictly fiction, but, like many extended case-studies and other accounts of psychotherapeutic encounters, they read like short-stories. Later on I will say more about the connection between the existential paradigm and indirect means of communication in general, and in the last section between the existential paradigm and narrative in particular.
5. Their mouthpiece is *Existential Analysis*.
6. I am only giving the briefest details here, but will return to these areas and others later in this chapter.
7. By "ultimate concerns," as we shall see, he means things like death and freedom.
8. See Sartre (1966, Parts 1 and 4).
9. A price existentialism pays for being paradigmatic is Popper's question of falsifiability. If, within the realm of conscious processes, *everything* can be interpreted existentially and there are no empirically refutable predictions that it can make, then it loses any claim it has to being scientific. (See Popper, 1963). However, and along lines similar to Freud's defense of psychoanalysis, if existential therapy was ineffective and if the probings of existential texts *persistently* did not accord with intuitions then these would be good grounds for doubt.
10. For a brief account of these concerns see the prologue to *Love's Executioner*.
11. See, e.g. (Becker 1973; Brown, 1959/1968; Rank, 1924/1929, 1941/1958; Tillich, 1952/1962).
12. The literal translation of the German *Unheimlichkeit* (see Heidegger, 1990/1926 Part 1, Section VI). Like comparing interpretations of the Oedipus myth, reading Freud's essay *The Uncanny* alongside

Heidegger's analysis is a useful heuristic for understanding the difference between existential and psychoanalytic orientations. (Freud, 1962, Vol. 17)

13. Faith he sees as a home of sorts, but there is some ambiguity regarding its mutability, and it's hard won to say the least.
14. For pointers towards empirical research that supports this, see Bee, (1998, pp. 514–517)
15. For instance, see Sartre's colorful example of the masochist paying to be whipped (1943/1966, p. 493).
16. I am reminded here of Ken Loach's approach to film making. A technique of his to encourage actors to identify with and care about their characters is not to let them know the entire plot in advance. In the Spanish Civil War story *Land and Freedom*, for instance, the cast were only given their lines on the day of shooting, before which they didn't know whether their character would make it through the day in one piece. (Source—The Late Show, BBC 2, May 15th 1995)
17. For a summary of the "characteristics of a mature, need-free relationship," see Yalom (1980, p. 373).
18. Daniel Dennett's term (cited in Flanagan, 1996, p. 69).
19. William James' expression, which Yalom draws our attention to. In *The Principles of Psychology* James lists five kinds of decision, the spirit of which is very close to Farber. (See James, 1890, Ch. 26, and Yalom, 1980, p. 315).
20. For interesting philosophical discussions and further references on character and moral psychology see Flanagan (1991).
21. "Fear of death determines the element of anxiety in every fear." (1962, pp. 46–47).
22. "From both my personal and professional experience, I have come to believe that the fear of death is always greatest in those who feel that they have not lived their life fully. A working formula is: the more unlived life, or unrealized potential, the greater one's death anxiety." (Yalom, 1989/1991, p. 111)
23. But not necessarily that close. For an extremely interesting (and ambitious) attempt to integrate various stage theories in psychology and philosophy, see Wilber (2000).
24. Relevant also here are Rank's three stages in the development of the will—"counter will," "positive will" and "creative will" (Yalom, 1980, pp. 295–297).
25. For example, Kierkegaard (1849/1989); Heidegger on our "they-self" (1990, Division I, Section IV and Division II, Section IV); Sartre on bad faith (1966, Part One, Ch.2; 1961), Camus (1951/1984); Tillich (1952/1962).
26. Interestingly, Kierkegaard, pre-empting Freud said "when insanity has a mental basis, it is always due to a hardening at some point in the unconscious" (1987, Vol. I p. 83).
27. See Nietzsche (1882/1974), Laing (1968, 1969), Hesse (1989), James (1890, 1915/1967), Tillich (1962), Kundera (1985).
28. "Style" has a distinctively Nietzschean meaning here by the way.
29. Also in the research tradition of social psychology the tendency to distort our memories in order to maintain a positive self-image has been thoroughly investigated under the banner of "cognitive dissonance" (Festinger, 1957). The theory states that instances where there is a perceived conflict between one's attitude to oneself and how one has in fact behaved creates a tension or anxiety that motivates an unconscious distortion of the recalled meaning of one's action. The types of (often implicit) self-beliefs revealed by these experiments include "I am not a hypocrite, "I say what I believe" and "I do what I want to do". (See Brown, 1965/1967, Ch. 11))
30. For a thorough treatment of this (though different in some respects to what my approach has been) see Flanagan (1991, Ch. 15).
31. See, for example, Alloy and Abramson (1982).
32. For example see the appendices of Zuckerman (1979).
33. Some widely chosen examples are Kierkegaard (1985), Conrad (1973), May (1991), and the relevance of "the voyage" to Laing's work (see Heaton, 1995).
34. See, for example, Herrigel (1953/1985).
35. For a sophisticated analysis of criteria see MacIntyre (1981) on "practices."

36. I don't think it's too much of a liberty to extend Nagel's analysis in this way. From an objective point of view we intellectually note this oddity, but subjectively we are emotionally disturbed by it. Filled up with, surrounded by a "meaningful" life we are aggrieved in some way by what objectivity reveals. Perhaps a sense of "absurdity" arrives at rare and fleeting moments where we bridge these points of view, but otherwise Nagel's point is nicely demonstrated simply by our difficulty in finding a single word or experience that captures what he has in mind.
37. I find Yalom too optimistic in this regard, so too Boss (for example see Boss, 1957/1963, Ch. 16).
38. See, for example, Brehm (1956). Similar motivations no doubt contribute to a conclusion reached from research into typical biases in decision making and judgment. One writer (Plous, 1993) concludes that the one best recommendation for debiasing our decision making processes is the "consideration of the alternative perspectives" (p. 256). This might sound like a very avoidable error, but the findings are that it is one we are heavily prone to.
39. A similar idea is discussed in moral philosophy under "agent regret" (see Rorty, 1980).
40. Marcia's (1980) development of Erikson's adolescent stage is sensitive to this.
41. Heidegger places a strong emphasis on guilt (or debt) in this sense as well, as do Sartre and Camus in their own ways, particularly with regard to our relations to others (rather than our past). See, for example, Camus (1984), and comments such as "The original sin is my upsurge in a world where there are others." (Sartre, 1966, p. 531).
42. Gordon Allport's expression (cited in Yalom, 1980, p. 431)
43. For similar, but ostensibly non-existential, accounts of the nature of emotions see Oatley and Jenkins (1996, Ch. 4).
44. For summary and further references see Oatley and Jenkins (1996, Chapters 1 and 4).
45. See Ratcliffe (2001).
46. Relevant to many of his works, but see especially Concluding Unscientific Postscript.
47. Conscious self-depreciation is a feature of both Yalom's and van Deurzen-Smith's writings. It is especially true of the former's popularist books like *Love's Executioner* and *Momma and the Meaning of Life*, and with the latter there are good examples in the Prologue to *Everyday Mysteries* and in the case study at the end of the book (especially pp. 271–273).
48. For example, "disidentification" (Yalom, 1980, p. 164) and his list of "artificial aids" such as the "existential shock" technique, "anticipatory regret" and the "tombstone exercise" for developing more authentic relations with death (Yalom, 1980, pp. 173–187, 1996/1997, pp. 179–180, pp. 270, 362). Perhaps one of the first such techniques was Nietzsche's eternal return: As a test of authenticity Nietzsche's demon—our therapist—asks, would we want to relive this life in all its details 'times without number'? (Nietzsche, 1974, p. 341)
49. Stephen Mulhall's summary of the meaning of anxiety in Heidegger (Mulhall, 1996).
50. van Deurzen-Smith says that the "client" must been seen as "his or her own source of light" (1997, p. 188) and that the analyst should "venture into ... the other's world experience as if they were going into unknown territory" (p. 218).
51. Paraphrasing rank.

REFERENCES

Alloy, L. B., & Abramson, L. Y. (1982). Learned helplessness, depression and the illusion of control. *Journal of Personality and Social Psychology*, *42*(6), 1114–1126.

Becker, E. (1973). *The denial of death*. New York: The Free Press.

Bee, H. (1998). *Lifespan development*. New York: Longman.

Boss, M. (1963). *Psychoanalysis and daseinsanalysis*. New York: Basic Books. (Originally published 1957)

Brehm, J. (1956). Post-decisional changes in desirability of alternatives. *Journal of Abnormal and Social Psychology*, *52*, 384–389.

Brown, R. (1967). *Social psychology*. London: Collier-MacMillan. (Originally published 1965)

Brown, N. O. (1968). *Life against death*. London: Sphere Books. (Originally published 1959)

Buber, M. (1947). *Between man and man*. London: Collins/Fontana.

Buber, M. (1970). *I and thou*. New York: Scribner. (Originally published 1922)

Camus, A. (1975). *The myth of Sisyphus*. London: Penguin. (Originally published 1941)

Camus, A. (1984). *The fall*. Harmondsworth: Penguin. (Originally published 1951)

Conrad, J. (1973). *Heart of darkness*. London: Penguin. (Originally published 1901)

Cooper, D. E. (1990). *Existentialism: A reconstruction*. Oxford: Blackwell.

Damasio, A. (1994). *Descartes' error*. New York: Putnam.

Damasio, A. (2000). *The feeling of what happens*. London: Vintage. (Originally published 1999)

de Beauvoir, S. (1994). *The ethics of ambiguity*. New York: Citadel Press. (Originally published 1948)

de Sousa, R. (1980). The rationality of emotions. In O. Rorty (Ed.), *Explaining emotions*. Berkeley, CA: University of California Press.

de Unamuno, M. (1954). *The tragic sense of life*. New York: Dover. (Originally published 1912)

Erikson, E. (1950). *Childhood and society*. New York: Norton.

Farber, L. (1966). *The ways of the will*. New York: Basic Books.

Festinger, L. (1957). *A theory of cognitive dissonance*. New York: Harper and Row.

Flanagan, O. (1991). *Varieties of moral personality*. Cambridge, MA: Harvard University Press.

Flanagan, O. (1996). *Self expressions*. Oxford: Oxford University Press.

Frankl, V. (1984). *Man's search for meaning*. New York: Washington Square Press. (Originally published 1959)

Freud, S. (1962). *The complete psychological works* (Vol. 17). London: Hogarth.

Friedman, M. (1989). *The healing dialogue in psychotherapy*. Northvale NJ: Aronson.

Fromm, E. (1957). *The art of loving*. London: George Allen and Unwin.

Gould, R. (1978). *Transformations: Growth and change in adult life*. New York: Simon and Schuster.

Heaton, J. M. (1995). The self, the divided self and the other. *Journal of the Society for Existential Analysis*, *6*(1), 31–60.

Heidegger, M. (1990). *Being and time*. Oxford: Blackwell. (Originally published 1926)

Herrigel, E. (1985). *Zen in the art of archery*. London: Penguin/Arkana. (Originally published 1953)

Hesse, H. (1989). *My belief*. London: Triad/Paladin.

Hjelle, L. A., & Ziegler, D. J. (1981). *Personality theories: Basic assumptions, research and applications*. New York: McGraw-Hill.

Hoeller, K. (Ed.). (1990). *Readings in Existential Psychology and Psychiatry*. Seattle: Review of Existential Psychology and Psychiatry.

James, W. (1890). *The principles of psychology*. New York: Henry Holt.

Jaspers, K. (1963). *General psychopathology*. Chicago: Regnery. (Originally published 1913)

Kierkegaard, S. (1980). The concept of anxiety. Princeton: Princeton University Press. (Originally published 1844)

Kierkegaard, S. (1985). *Fear and trembling*. London: Penguin. (Originally published 1843)

Kierkegaard, S. (1987. *Either/or*. Princeton: Princeton University Press. (Originally published 1842)

Kierkegaard, S. (1989). *The sickness unto death*. London: Penguin. (Originally published 1849)

Kierkegaard, S. (1992). *Concluding unscientific postscript*. Princeton: Princeton University Press. (Originally published 1846)

Kundera, M. (1985). The unbearable lightness of being. London: Faber and Faber.

Laing, R. D. (1968). *The divided self*. Harmondsworth: Penguin.

Laing, R. D. (1969). *Self and others*. Harmondsworth: Penguin.

Levinson, D. (1978). *The seasons of a man's life*. New York: Knopf.

MacIntyre, A. (1981). *After virtue*. London: Duckworth.

Marcel, G. (1949). *Being and having*. Westminster, UK: Dacre Press.

Marcel, G. (1951). *Homo viator*. New York: Harper and Brothers.

Marcia, J. (1980). Identity in adolescence. In J. Adelson (Ed.), *Handbook of adolescent psychology*. New York: Wiley.

Maslow, A. (1987). Motivation and personality. New York: Harper and Row. (Originally published 1954)

Merleau-Ponty, M. (1979). The phenomenology of perception. London: Routledge and Kegan Paul. (Originally published 1945)

May, R. (1991). *The cry for myth*. New York: Norton.

May, R., Angel, E., & Ellenberger, H. (Eds.). (1958). *Existence: A new dimension in psychiatry and psychology*. New York: Basic Books.

Mischel, W. (1968). *Personality and assessment*. New York: Wiley.

Mulhall, S. (1996). *Heidegger and being and time*. London: Routledge.

Nagel, T. (1979). *Mortal questions*. Cambridge, UK: Cambridge University Press.

Nagel, T. (1986). *The view from nowhere*. Oxford: Oxford University Press.

Nehamas, A. (1985). *Nietzsche: Life as literature*. Cambridge, MA: Harvard University Press.

Nietzsche, F. (1968). *The twilight of the idols*. Harmondsworth: Penguin.

Nietzsche, F. (1974). *The gay science*. London: Random House. (Originally published 1882)

Nussbaum, M. (1990). *Love's knowledge*. Oxford: Oxford University Press.

Nussbaum, M. (2002). *Upheavals of thought: The intelligence of emotions*. Cambridge, NY: Cambridge University Press.

Oatley, K., & Jenkins, J. (1996). *Understanding emotions*. Oxford: Blackwell.

Plous, S. (1993). *The psychology of judgement and decision making*. New York: McGraw-Hill.

Popper, K. (1963). *The logic of scientific discovery*. New York: Basic Books.

Raabe, P. B. (2001). *Philosophical counselling: Theory and practice*. Westport, Connecticut: Praeger.

Rank, O. (1929). *The trauma of birth*. New York: Harcourt Brace. (Originally published 1924)

Rank, O. (1958). *Beyond psychology*. New York: Dover. (Originally published 1941)

Ratcliffe, M. (2002). Heidegger's attunement and the neuropsychology of emotion. *Phenomenology and the Cognitive Sciences*, *1*(3), 287–312.

Rogers, C. (1967). On becoming a person. London: Constable. (Originally published 1961)

Rorty, R. (1989). *Contingency, irony and solidarity*. Cambridge: Cambridge University Press.

Rorty, A. O. (Ed.). (1980). Agent regret. In *Explaining Emotions*. Berkeley, CA: University of California Press.

Sartre, J.-P. (1961). *The age of reason*. Harmondsworth, Penguin. (Originally published 1947)

Sartre, J.-P. (1966). *Being and nothingness*. New York: Washington Square Press.

Sartre, J.-P. (1989). 'Dirty Hands' in *No Exit and Three Other Plays*. New York, Vintage. (Originally published 1947)

Sartre, J.-P. (1992). *Notebooks for an ethics*. Chicago: University of Chicago Press. (Originally published 1983)

Solomon, R. (1993) *The passions*. Indianapolis: Hackett. (Originally published 1976)

Taylor, S., & Brown, J. (1988). Illusion and well being: A social psychological perspective on mental health. *Psychological Bulletin*, *3*(2), 193–210.

Tillich, P. (1962). *The courage to be*. London: Fontana Library. (Originally published 1952)

van Deurzen-Smith, E. (1997). *Everyday mysteries*. London: Routledge.

van Deurzen-Smith, E. (1999). Existentialism and existential psychotherapy. In C. Mace (Ed.), *Heart and soul: The therapeutic face of philosophy*. New York: The Guilford Press.

van Kaam, A. (1961). Clinical Implications of Heidegger's Concept of Will, Decision and Responsibility. *Review of existential psychology and Psychiatry, 1*, 205–216.

van Kaam, A. (1990). Existential psychology as a comprehensive theory of personality. In K. Hoeller (Ed.), *Readings in existential psychology and psychiatry*.

Wheway, J. (1999). The dialogical heart of intersubjectivity. In C. Mace (Ed.), *Heart and soul: The therapeutic face of philosophy*. London: Routledge.

Wilber, K. (2000). *Integral psychology*. Boston: Shambhala.

Woolf, V. (1977). *The waves*. London: Grafton.

Yalom, I. D. (1980). *Existential psychotherapy*. New York: Basic Books.
Yalom, I. D. (1991). *Love's executioner*. London: Penguin. (Originally published 1989)
Yalom, I. D. (1992). *When Nietzsche wept*. New York: HarperCollins.
Yalom, I. D. (1997). *Lying on the couch*. New York: Harper Perennial. (Originally published 1996)
Yalom, I. D. (1999). *Momma and the meaning of life*. London: Piatkus.
Yeats, W. B. (1974). *Selected poems*. London: Pan.
Zuckerman, M. (1979). Attribution of success and failure revisited, or: The motivational bias is alive and well in attribution theory. *Journal of Personality*, *47*, 245–287.

CHAPTER 10

CONCLUSION: PHENOMENOLOGY AND PSYCHOLOGICAL SCIENCE

PETER D. ASHWORTH and MAN CHEUNG CHUNG

The direct impact of phenomenological thinking on psychological science has not been great (as we have seen in Chapter 2). There has certainly been indirect influence, especially through the migration of German Gestalt psychologists to the United States during the 1930s. There has also been a definite effect of existential phenomenology—the engagement of phenomenology with the elements of Kierkegaard's existentialist thinking, engineered by Heidegger. This had an undeniable role in the establishment of humanistic psychology, some kinds of counseling, and (perhaps more peripherally) on psychiatry. But the authors of this book are united in their view that the low impact of phenomenological thinking on psychology is much to be regretted. Psychologists have in very large measure chosen a natural science model of research somewhat in line with the positivist tendency (though not always fully, technically positivist, since the criterion of truth is not always that the "facts" should be observable and the relationships between observables). It is, however, true to say that psychological science generally assumes—without reflecting overmuch on the assumption—that there is one real world which has determinate characteristics, and the purpose of science is to model this world in its theories. These theories will show how certain variables interrelate, especially how they relate to each other in a cause-and-effect fashion. Mathematical formulations of the relationships between variables are to be sought if at all possible. The purpose of research is to test hypotheses regarding relationships between variables, and to reach, by closer and closer approximation, theories which can begin to be regarded as having the status of scientific laws.

Phenomenologists see this approach as a mistake. A very different line of research is possible. In this chapter, we will sketch out what it might look like, drawing on the work of pioneer phenomenological psychologists such as the Giorgis (Chapters 3 and 4).

But first we must address the historical and philosophical question of whether it is right to label any line of research within an empirical science "phenomenological psychology." What is phenomenological psychology, and what relation does it bear to psychology as generally practiced?

Strictly we should be referring to *phenomenologically based psychology* here rather than phenomenological psychology. The founder of modern phenomenology, Edmund Husserl (1859–1938) aimed to establish a large set of philosophical disciplines, each having the purpose of clarifying the basic conceptual structure of a science or scholarly field. So, as well as a phenomenological psychology there would be a phenomenological biology, a phenomenological jurisprudence, a phenomenological theology, a phenomenological geography, etc. Phenomenological psychology, then, would not be empirical in the usual sense, but would be a philosophical field. The way in which a discipline's foundations would be established phenomenologically would be by reflecting on the fundamental experiences with which the discipline seems to be concerned (maybe, for the arbitrary list just given, reflection would be turned to living things, justice, fundamental meaning, and spatial structuring). Phenomenological psychology in Husserl's sense, then, would be a reflective discipline aimed at clarifying the concepts which seem to be fundamental to psychology, and reflecting on them in a special way—attending to the way the things referred to by the concepts appear in experience. So, if motivation is a fundamental concept of psychology, part of phenomenological psychology would be to describe instances of motivation *as they appear in awareness*. And clarification of the structure of conscious experience as such would be part of the groundwork needed before a *particular* kind of experience—such as a motivational one—could be described adequately.

Husserl went some way to founding a phenomenological psychology of this sort (e.g., Husserl, 1925/1977). His philosophical work on the basic concepts of psychology is certainly not without interest for the empirical science, but generally psychologists concerned with phenomenology have not followed Husserl's way. Rather than trying to establish the "essential structures" of the discipline, they have turned attention to particular areas of experience and have sought to draw out the human meanings of these areas, using empirical materials such as interviews and observation. It is a phenomenologically based empirical effort. For short, we will call the empirical work in psychology which is based on phenomenology, "phenomenological psychology."

Though deviating from what Husserl meant by phenomenological psychology, such work is nevertheless phenomenologically based because it takes as its focus (in a way that no other trend in psychology does) *the person's experience as such*. Whether the person is right or wrong, in tune with the actual evidence of reality or not, the aim is to describe the experience of just as it is experienced, "in its appearing."

The claim that phenomenological psychology is distinct from all the rest of psychology in attending to experience in the way it does may seem wild. After all, questionnaires and other self-report instruments ask for accounts of experience; psychophysics asks its participants to report on stimuli; clinical psychology often asks for—and takes very seriously—people's accounts of their thoughts and feelings, and so on. Concern with experience is not unusual, apparently, in psychology. All this is true. What is distinctive about the phenomenological approach, however is the strictness with which *all reference to reality is set aside*. This is the so-called epochē or "bracketing." It is a fundamental concept for any phenomenologically based work because it points to the locus of all relevant research. That locus is nothing but the field of awareness. Reality is bracketed and, in the selfsame move, attention is turned to experience alone (we have here the "phenomenological reduction"). Work is carried out "within the reduction" (Husserl, 1913/1983). The following comment of Husserl's, explaining that the "real world" is to be bracketed in order to focus on experience as such, describes well the position of phenomenologically based psychology:

> ... I am not negating this "world" as though I were a sophist; I am not doubting its factual being as though I were a skeptic; rather I am exercising the "phenomenological" epochē which also completely shuts me off from any judgment about spatiotemporal factual being. (Husserl, 1913/1983, §32, p. 61)

In practice, no phenomenologically based research in psychology dissents from this. At this level, the epochē might be seen as the mirror-image of Skinner's decisive bracketing of consciousness (in the sense of inner personal awareness). Skinner (e.g., 1993) choosing to regard the world of natural science as the sole reality, and regarding the organism as an intrinsic part of the spatiotemporal factual world, sets aside conscious experience:

> Our increasing knowledge of the control exerted by the environment ... makes it possible to interpret a wide range of mental expressions .... Some can be "translated into behavior," others discarded as unnecessary or meaningless. (Skinner, 1993, p. 17)

In an exactly opposite move, phenomenological psychology brackets such "objective" reality and focuses on conscious experience.

INTENTIONALITY

Turning from objective reality to experience as such, we discover intentionality. But there is a history of debate concerning how this is to be characterized, and as we shall see, this question is of central interest to a phenomenological psychology.

> Every mental phenomenon includes something as an object within itself, although they do not all do so in the same way. In presentation something is presented, in

judgment something is affirmed or denied, in love loved, in hatred hated, in desire desired and so on. (Brentano 1874/1995, p. 88)

It is clear that, by the phrase "mental phenomenon," Brentano meant only the conscious phenomenon. And later Brentano would substitute "reference to an object" for "something as an object within itself." However this may be, his meaning is fairly plain. Brentano (1874/1995) regarded immediate experience as a process or *act*, and different kinds of experience are to be distinguished by the particular way in which consciousness relates to the object of experience. Judgment and perception, for instance, each involve a different orientation to the object. The definitive feature of conscious activity for Brentano was, in a word, its *intentionality.* The fact that consciousness is intentional was definitive. "All consciousness is consciousness of something". And psychology had the task of delineating the various ways in which consciousness could relate to its objects.

As we saw in Chapter 2, Husserl (1913/1983) refined Brentano's statement of intentionality, for Husserl believed it could be understood as picturing the world as divided into the "outer" reality and its "inner" correlate.

> ...[I]t should be well heeded that *here we are not speaking of a relation between some* psychological occurrence—called a mental process—and another real factual existence—called an object—nor of a *psychological connection* taking place *in Objective actuality* between one and the other. Rather we are speaking of mental process purely ldots (Husserl, 1913/1983, §36, p. 73)

To render this point unmistakable, a particular understanding of the idea of intentionality is required:

> ...[N]othing is accomplished by saying and discerning that every objectivity relates to something objectivities, that every judging relates to something judged, etc.... For without having seized upon the peculiar own ness of the transcendental attitude and having actually appropriated the pure phenomenological basis, one may of course use the word, phenomenology; but one does not have the matter itself.... (Husserl, 1913/1983, §87 p. 211)

Husserl is clear here that both the "mode of consciousness" and the "object of this consciousness" are (as it were) *mine*, they are both within personal experience or awareness. So "the phenomenon" is what appears (the intentional object) which is described precisely as it appears, with full attention also given to the conscious mode (perception, for example) in which it appears. So there are two "aspects" of the phenomenon as described by Husserl (1913/1983), the *noema*, the *object* of awareness, and the *manner* in which one is aware of it, the *noesis*.

So Merleau-Ponty was absolutely correct in his assertion:

> We can now consider the notion of intentionality, too often cited as the main discovery of phenomenology, whereas it is understandable only through the reduction. (Merleau-Ponty, 1945/1962, p. xviii)

In sum, turning attention to experience as such, just as given, bracketing the issue of the "reality" of its objects, we discover that all consciousness is intentional. For each noesis there is a noema.

Heidegger (1975/1988—actual date of the work, 1932) developed an exceedingly important elaboration of Husserl's version of intentionality, which is of great consequence for phenomenological psychology. The importance lies both in its implications for methodology and in the effect it has on the scope of the actual descriptions achieved by phenomenological psychologists.

Heidegger urges us to consider the conscious subject (the meaning, roughly, of his term "the Dasein") in their actual existence. The being of the subject can certainly be approached through a consideration of intentionality:

> ... the Dasein's comportments have an *intentional character* and ... on the basis of this intentionality the subject already stands in relation to things that it itself is not. (Heidegger, 1975/1988, §15, p. 155)

Something needs to be said about this statement. Heidegger is purposely moving the discussion well away from any imagery distinguishing "the mental world" from "the material world" (in a way, completing the efforts in this direction of Husserl). The use of the notion of comportment is tied in with the idea of the Dasein as "being-in-the-world" in the sense of being already, before everything, a participant in the world—physical actions or mental acts are within and directed to the world. Comportment, then, is *intentional.*

If we are to say that "all comportment is within and towards a world," as a new Heideggerian expression of intentionality founded on being-in-the-world (see Heidegger, 1975/1988, p. 161, §15c), then another aspect of intentionality comes into view. This is critical. It is the inevitable *subjectivity* of the world—my world:

> The surrounding world is different in a certain way for each of us, and notwithstanding that we move in a common world. (Heidegger, 1975/1988, p. 164, §15c).

Heidegger is emphatic that the fact of the subjectivity of the world shatters the previous notion of intentionality:

> One things is certain. The appeal to the intentionality of comportments toward things does not make comprehensible the phenomenon occupying us, or, speaking more cautiously, *the sole characterization of intentionality hitherto customary in phenomenology proves to be inadequate and external.* (Heidegger, 1975/1988, p. 161, §15c).

Now, it is not certain that Heidegger is entirely right in criticizing Husserl here, which he does, if only by implication. First of all, Heidegger is—and this must be emphasized—still thinking *within the reduction*. He is a transcendental philosopher. Without specifically mentioning it, he is concerned with the world seen from the point of view of experience. He is not taking an external stance. Secondly, Husserl himself stated very clearly that, within the reduction, we are speaking about *my* world. So the subjectivity of the world, if it is being underlined by Heidegger here, it is still an elaboration or explication of Husserl's insight that the world revealed in intentionality has a "peculiar ownness.'

Heidegger develops the phenomenology of intentionality by utilizing an old distinction in the theory of the self (it appears, for example, in William James, 1890/1950; see Chapter 2 of the present volume). There is a chasm set between two different meanings of self: (a) the "anonymous" self which is the subject of consciousness and (b) the self as an object of reflective awareness, the self as characterized.

Husserl was clear that the self-as-subject has no characteristics, there is no essentialism of self:

> [T]he Ego living in mental processes ... is not something taken *for itself* and which can be made into an Object *proper* of an investigation. Aside from its "modes of relation" or "modes of comportment" the Ego is completely empty of essence-components, has no explicable content, is indescribable in and for itself: it is a pure Ego and nothing more. (Husserl, 1913/1983, §80, p. 190)

For Heideggger, too, there is a primary, pre-reflective "understanding" of self—in the sense of subjectness, and this is without characteristics. Self-characterization is made possible by reflection on comportment and world:

> [The Dasein] never finds itself otherwise than in the things themselves, and in fact in those things that daily surround it. It finds *itself* primarily and constantly *in things* because, tending them, distressed by them, it always somehow rests in things. Each of us is what he pursues and cares for. In everyday terms, we understand ourselves and our existence by way of the activities we pursue and the things we take care of. (Heidegger, 1975/1988, page 159, §15).

It seems that Heidegger wants to say that to exist in the human way (Dasein) is to already find oneself as within the structure of meaning designated by "world." In intentionality, therefore, we find *our selves* in the subjective world to which our comportment is directed. Any phenomenological psychology will expect to see subjectivity in an experience of a particular phenomenon.

THE REDUCTION: THE EPOCHĒ

"*Will expect to see*?" But in entering into the reduction, and exercising the epochē, have we not bracketed expectations? It has been emphasized that Husserl and Heidegger are transcendental philosophers. They both turn attention to "the things themselves" as given in experience, describing them just as they appear. This turn to conscious experience is the phenomenological reduction, and in performing it all tendencies to locate experience within the spatio-temporal "real" world are subjected to a stoppage, an epochē, a bracketing. But is it also pointed out, notably within phenomenological psychology, that presuppositions and expectations must also be subjected to an epochē. The matter of experience must be described as it is, without the researcher importing their prior experience, their interpretations, their knowledge of scholarly and scientific views of the phenomenon (Ashworth, 1996).

With the new formulation of intentionality which Heidegger puts forward, we find that experience is embedded in a world which already speaks of the individual perspective of the experiencer. (This is so whether the experiencer under investigation is the researcher themselves or another person.) Now this does not hazard the fundamental sense of Husserl's epochē—the setting aside of reality in order to focus on the description of the phenomenon as experienced. But the attempt to "cleanse" the description of experience of all accretions is not (fully) possible. Beyond the basic phenomenological reduction, the bracketing of further theoretical and other presuppositions (etc.) by the psychologist may be seen as a controversial technical move in empirical work. Where it is used, it is intended simply to keep the research firmly in the realm of experience. It is to be emphasized that the key characteristic of phenomenological psychology is its *openness to the research participant's experience*. Renunciation of our own prior understandings, where it is carried out, is in the service of entering the experience (of oneself or another) and openness to what appears. It ought also to be said that *reflexivity* on the part of researchers, such that they are sensitive to the assumptions, presuppositions, linguistic formulations and so on, which they bring to awareness in the effort to enter or remain within the realm of the reduction, may well provide material which is itself worth scrutiny in wrestling with the meaning of the phenomenon (Finlay & Gough, 2003).

Phenomenological psychologists take a wide range of positions on this issue. Giorgi (1985) takes a relatively Husserlian line in his phenomenological psychology, engaging in quite minute interrogation of the text of an interview and tying descriptions of experience to this evidence. Nevertheless, in the end, the account is a "psychological" one—and this perspective might very well not be that originally entertained by the research participant. Elucidation, for Giorgi, entails definite empathic work by the researcher, though still always "descriptive." Smith (2004) makes it plain that his phenomenology is "interpretive." This is despite the fact that his work, which is largely in health psychology, is truly phenomenological being located precisely with experience as such. Though he wishes to describe the research participant's experience purely in their own terms, nevertheless he emphasizes that there is an interpretive element. His analyses of experience are set up in order to shed light on his own area of research interest, and he allows himself to interrogate interview text and draw out allusions and metaphors which the patient might well not have noted (though they were present at least by implication in their own talk). Van Manen (1990, 1991) might go further than Smith in bringing metaphors and other linguistic forms to bear on the elucidation of experience. Here there is no doubt that the account of the person's experience goes much beyond what they would be able to say themselves. Nevertheless, we still are shown work *within the reduction*, work aimed at elucidating and clarifying experience. Bringing the unstated taken-for-granted implications of accounts of experience to expression in the analyses of phenomenological psychology is certainly within the meaning of the reduction. As Husserl wrote:

> ... [W]hat is decisive consists of the absolutely faithful description of what is actually present in phenomenological purity and in keeping at a distance all the interpretations transcending the given. (Husserl, 1913/1983, §90, p. 218)

Interpretation and Discourse

"The given" is not simply the obvious. But discipline in interpreting is necessary. Ricoeur (1970) makes a valuable distinction here. In discussing Freud, he writes of the *hermeneutics of suspicion*, pointing out that interpretations of the Freudian kind bring into play a framework of understanding that is specifically *not* in the analysand's awareness. Such interpretations "transcend the given" hugely. Phenomenologically based psychological work is *never* carried out in terms of such a meaning of interpretation. Ricouer contrasts this with what he calls the *hermeneutics of meaning-recollection* in which whatever interpretive work is done is in the service of "bringing out" the meaning of the experience. This kind of interpretation is indubitably phenomenological.

There is no question but that, as a qualitative methodology, phenomenological psychology contrasts with psychologies of the "discursive turn." The work of phenomenologists often rests on language, of course. The interviewee "expresses" her experience; the researcher reads this expression in terms of socially-available constructs. Nothing is clearer. The words and the experience—there is a great distinction here between phenomenological and discursive approaches. For, though there is no resistance at all among phenomenologists to the fundamental experience-shaping function of language and discursive practices, the strongly-held position of phenomenology is that it is *of experience that language speaks* (Merleau-Ponty, 1945/1962). And, it is often the case that phenomenological psychology needs to stretch the meaning of words or invent a new vocabulary in order to draw attention to the nuances of experience. Further than this, researchers find that, when research participants begin to struggle for words, or multiply examples, or begin to lose clarity—it is precisely at this point that discoveries concerning the meaning of a phenomenon for them are near.

Phenomenological Psychology and the Wider Discipline of Psychology

Phenomenological psychology is exclusively and wholly absorbed in the elucidation of experience. What a renunciation this entails! It is arguable that the unspoken aim of much undergraduate psychology teaching is to inculcate the student into a world consisting of *variables* that interact in cause-and-effect ways such that behavior and experience become explicable or open to experimentation. Conscious experience, if it is considered to be relevant, is to be regarded as the lawful outcome of factors in the world such as physiological conditions and objective events. Without denying or accepting these ways of portraying the psychological, phenomenological psychology brackets it all, for Husserl's call to turn towards *the things themselves* is at the same time a call to turn *away* from the assumptions of contemporary psychological research. It would no doubt be possible to take the findings of a phenomenological study and subject them to the usual techniques of a quantitative psychology. Maybe a questionnaire could be devised on the basis of certain findings of phenomenological psychology, and used in studies yielding quantitative results. Maybe phenomenological findings could be developed into a measure and correlations with other "personal tendencies" could be investigated and individual differences mapped out. Maybe causal factors could be investigated. But this would no longer be phenomenological psychology—it would entail an exit from the reduction. It would be making positive assumptions about the spatiotemporal, "objective" world, which would be taken as fundamentally the world described in the terms of a physical science, and open to mathematicisation. Phenomenological psychology does not operate in terms of this view of the world but in terms of a world in which experience is primary and physical science is a derivative account (Merleau-Ponty, 1945/1962).

My world is one of personal meanings—and the underlying value of phenomenological psychology is to empower its practitioners and readers to re-assert the human perspective.

REFERENCES

Ashworth, P. D. (1996). Presuppose nothing! The suspension of assumptions in phenomenological psychological methodology. *Journal of Phenomenological Psychology, 27*, 1–25.

Brentano, F. (1995). *Psychology from an empirical standpoint.* London: Routledge. (Originally published 1874)

Finlay, L. & Gough, B. (Eds.). (2003). *Doing reflexivity: Critical illustrations for health and social science.* Oxford: Blackwell.

Giorgi, A. (1985). Sketch of a psychological phenomenological method. In A. Giorgi (Ed.), *Phenomenology and psychological research.* Pittsburgh: Duquesne University Press.

Heidegger, M. (1982 originally published 1975*). *The basic problems of phenomenology.* Bloomington: Indiana University Press. [*1927 is the date of the original lecture series]

Husserl, E. (1977). *Phenomenological psychology.* The Hague: Martinus Nijhoff. (Originally published 1925).

Husserl, E. (1983). *Ideas pertaining to a pure phenomenology and a phenomenological philosophy.* First Book. Dordrecht: Kluwer. (Originally published 1913).

James, W. (1950). *The Principles of psychology* (Vol. 1). New York: Dover. (Originally published 1890).

Merleau-Ponty, M. (1962). Phenomenology of perception. London: Routledge and Kegan Paul. (Originally published 1945)

Ricoeur, P. (1970). *Freud and philosophy: An essay on interpretation.* New Haven: Yale University Press.

Skinner, B.F. (1993). *About Behaviorism.* London: Penguin.

Smith, J.A. (2004). Reflecting on the development of interpretative phenomenological analysis. *Qualitative Research in Psychology, 1*, 39–54.

van Manen, M. (1990). *Researching lived experience.* New York: State University of New York Press.

van Manen, M. (1991). *The tact of teaching. The meaning of pedagogical thoughtfulness.* London, Ontario: The Althouse Press.

NAME INDEX

Abraham, 189
Abramson, L.Y., 192, 193
Adam, 186
Adler, A., 125
Alloy, L.B., 192, 193
Allport, G, 3, 14, 32, 33, 36, 41, 42, 61, 67, 193
Alzheimer, A., 151
American Psychiatric Association, 193
Angel, E., 170, 195
Angell, J., 60, 66, 67
Aristotle, 139, 148, 183
Arnold, M., 187
Arp, K., 113
Ashworth, P.D., 1, 2, 3, 4, 9, 11, 20, 38, 42, 92, 98, 100, 121, 197, 203, 205
Avenarius, R., 151, 167

Bach, J.S., 25, 26, 129
Barbone, S., 143
Barnes, H., 138, 142, 143
Baudelaire, C., 124
Bayle, A., 150, 151, 152
Beaufret, J., 130, 139, 140, 141, 143
Becker, E., 172, 191, 193
Bee, H., 192, 193
Beethoven, L, 25, 26
Benda, C., 143
Bengtsson, J., 91, 95, 100
Bernet, R., 102, 115, 116, 117, 118, 121
Binswanger, L., 8, 134, 141, 142, 143, 149, 150, 152, 153, 155, 159, 161, 170
Blankenburg, W., 142
Bleuler, E., 8, 152, 153, 167
Bleuler, M., 153
Blos, P., 139, 143
Blumenthal, A.L., 56, 67
Boehm, R., 139
Boring, E., 12, 42, 54, 59, 64, 66, 67
Boss, M., 7, 8, 38, 42, 134, 141, 142, 143, 147, 148, 149, 150, 151, 152, 153, 154, 155, 156, 157, 158, 159, 160, 161, 162, 163, 164, 165, 166, 167, 170, 193
Brehm, J., 193
Brentano, F., 13, 19, 42, 53, 58, 63, 64, 93, 200, 205
Brown, E.M., 150, 167
Brown, J., 182, 183, 195
Brown, N.O., 172, 175, 191, 194
Brown, R., 192, 194
Buber, M., 169, 172, 175, 177, 184, 194
Bugental, J.F.T., 36, 42
Bühler, C., 36, 42, 66

Camus, A., 172, 174, 185, 189, 190, 192, 193, 194
Canfield, F., 143
Cannon, B., 138, 142, 143
Carr, H., 58, 60, 66, 67
Catalano, J.S., 116, 121
Chessick, R., 143
Chestnut, W.J., 74, 87
Chung, M.C., 1, 3, 9, 197
Clément, C., 138, 143
Cloonan, T., 70, 87
Combs, A.W., 3, 35, 41, 42, 44, 70
Conrad, J., 192, 194
Contat, M., 139, 140, 143
Cooper, D.E., 173, 177, 191, 194

Corbin, H., 141
Cumming, R.D., 21, 42

Dahlberg, H., 92, 100
Dahlberg, K., 5, 6, 38, 42, 89, 92, 98, 100
Dahlstrom, D., 119, 120
Damasio, A., 187, 194
Danziger, K., 56, 67
Darwin, C., 61, 159
de Beauvoir, S., 26, 28, 42, 129, 139, 143, 174, 177, 194
de Sousa, R., 187, 194
de Unamuno, M., 177, 178, 194
Dennett, D., 192
Derrida, J., 41
Descartes, R., 106, 107, 112, 115, 116, 117, 118, 119, 123, 157
Dewey, J., 60
Dilthey, W., 105, 107, 158
Drew, N., 38, 42, 98, 100
Dreyfus, H., 111, 112, 113, 121
Drummond, J.J., 113

Ebbinghaus, H., 54, 67
Edie, J.M., 116, 120, 121
Ellenberger, H., 170, 195
Ellis, A., 178
Embree, L., 103, 104, 108, 113, 121, 122
Erikson, E., 124, 138, 139, 143, 181, 193, 194
Evans, J.C., 103
Eysenck, H., 32

Farber, L., 170, 172, 179, 181, 192, 194
Fechner, G., 12, 13, 42
Fell, J., 138, 143
Festinger, L., 192, 194
Fink, E., 100, 117, 118, 121
Finlay, L., 203, 205
Flanagan, O., 186, 187, 192, 194
Flaubert, G., 124
Føllesdal, D., 112, 121
Foucault, M., 150, 167, 170
Frankl, V., 170, 172, 183, 184, 194
Freud, S., 8, 25, 124, 125, 127, 131, 136, 138, 142, 147, 148, 152, 153, 155, 156, 158, 159, 160, 161, 164, 165, 166, 167, 168, 179, 186, 190, 191, 192, 194, 204
Friedman, M., 177, 194
Frijda, N., 187
Fromm, E., 172, 176, 194

Gadamer, H.-G., 6, 90, 92, 94, 95, 98, 99, 100
Gallanter, E, 16, 43
Genet, J., 138
Giorgi, A., 3, 4, 36, 37, 42, 45, 63, 67, 69, 70, 73, 87, 91, 92, 100, 198, 203, 205
Giorgi, B., 5, 69, 198
Goldfried, M.R., 74, 87
Gough, B., 203, 205
Gould, R., 170, 194
Grenness, C., 16, 17, 43
Griessinger, W., 150, 151
Groth, M., 7, 8, 123, 141, 142, 143
Guignon, C., 166, 168
Gurwitsch, A., 29, 30, 42

Halling, S., 38, 44
Hanly, C., 143
Hanscomb, S., 8, 9, 169
Harré, R., 32, 44
Hart, J., 110, 113, 117, 121
Hatab, L., 140, 143
Heaton, J.M., 192, 194
Hegel, G.W.F., 173
Heidbreder, E., 59, 61, 67
Heidegger, M, 3, 6, 7, 8, 9, 22, 24, 25, 26, 27, 28, 37, 38, 41, 42, 90, 94, 100, 101, 102, 103, 104, 105, 106, 107, 108, 109, 110, 111, 112, 113, 114, 115, 117, 118, 119, 120, 121, 123, 129, 130, 132, 133, 134, 135, 136, 137, 138, 139, 140, 141, 142, 143, 147, 148, 149, 150, 152, 154, 155, 156, 157, 159, 161, 162, 163, 164, 165, 166, 167, 168, 169, 170, 171, 173, 174, 175, 177, 181, 183, 185, 187, 191, 192, 193, 194, 197, 201, 202, 203, 205
Heider, F., 3, 31, 42
Heine, H., 155
Henley, T., 38, 43
Herrigel, E., 192, 194
Hesse, H., 181, 192, 194
Hitler, A., 148
Hjelle, L.A., 182, 194
Hoeller, K., 186, 189, 191, 194
Holderlin, F., 166
Hume, D., 159, 162
Humphrey, G., 64, 65, 66, 67
Husserl, E., 2, 3, 4, 5, 6, 7, 8, 9, 11, 13, 15, 17, 18, 19, 20, 21, 22, 23, 24, 25, 27, 28, 29, 33, 34, 37, 41, 42, 45, 46, 47, 48, 49, 50, 51, 52, 53, 56, 58, 59, 60, 63, 64, 65, 66, 67, 68, 69, 70, 74, 76, 78, 79, 83, 84, 87, 89, 90, 91, 92, 93, 94, 95, 96, 97, 98, 99, 100, 101, 102, 103,

104, 105, 106, 107, 108, 109, 110, 111, 112, 113, 114, 115, 116, 117, 118, 119, 120, 121, 124, 133, 150, 167, 198, 199, 200, 201, 202, 203, 204, 205
Huston, J., 124
Hycner, R., 178

Isaac, 189

James, W., 4, 12, 13, 14, 33, 34, 41, 43, 57, 61, 62, 63, 67, 68, 180, 181, 192, 194, 202, 205
Janzarik, W., 153, 168
Jaspers, K., 6, 101, 106, 107, 130, 141, 151, 153, 157, 158, 159, 165, 169, 168, 194
Jauregg, W, 153
Jenkins, J., 193, 195
Jenner, F.A., 8, 147, 150, 168, 170
Jung, C., 8, 125, 136, 147, 153, 154, 159, 160, 161

Kafka, F., 142
Kaiser, H., 178
Kant, I., 19, 150, 154, 156, 160, 166, 168, 177
Keen, E., 138, 144
Kendall, T., 150, 168
Kern, I., 102, 121
Kierkegaard, S, 24, 26, 28, 106, 139, 142, 169, 171, 174, 177, 181, 186, 188, 189, 192, 194, 197
Kisiel, T., 117, 119, 121
von Kleist, H., 160
Köhler, W., 29
Kraepelin, E., 148, 151, 152, 153, 159, 168
Kuhn, T.S., 155, 168
Kulpe, O., 58, 64, 65, 66, 68
Kundera, M., 181, 192, 194
Kvale, S., 16, 17, 43

La Fave, L., 143
Lacan, J., 142, 170
Laing, R.D., 26, 43, 170, 176, 181, 192, 194
Langenhove, L., 32, 44
Lapointe, F., 138, 139, 144
Leibniz, G.W., 115
Levinas, E., 108, 109, 110, 122
Levinson, D., 170, 181, 194
Lewin, K, 3, 31, 35, 43
Light, S., 139, 144
Linschoten, H., 62, 68
Lipps, H., 58
Loach, K., 192
Lohner, F., 141
Lucas, U., 92, 100, 121
Luckmann, T., 38, 43

Mabel, 186
MacDonald, P.S., 6, 7, 101, 102, 122
Mace, C., 195
MacIntyre, A., 192, 195
Macleod, R., 3, 34, 41, 43, 70
Macquarrie, J., 142, 157, 195
Mahrer, A.R., 74, 87
Mairet, P., 140
Mandler, G., 64, 65, 68
Mandler, J.M., 64, 65, 68
van Manen, M., 36, 37, 44, 203, 205
Marbach, E., 102, 121
Marcel, G., 130, 141, 170, 175, 177, 195
Marcia, J., 193, 195
Marx, K., 25, 124, 133, 139, 140, 173
Maslow, A., 36, 43, 176, 178, 180, 182, 195
May, R., 138, 170, 186, 188, 192, 195
Mayer, R., 65, 68
Mead, G.H., 14, 33
Meinong, A., 58, 64
Melle, U., 104
Merleau-Ponty, M., 1, 2, 6, 9, 26, 28, 38, 39, 40, 41, 43, 69, 70, 80, 88, 90, 91, 92, 94, 95, 97, 98, 99, 100, 101, 103, 117, 118, 122, 169, 174, 195, 200, 204, 205
Messer, A., 58
Miller, G.A., 16, 43, 85
Minkowsky, E., 170
Mischel, W., 176, 195
Misiak, H., 35, 36, 43
Mohanty, J.N., 53, 68
Moran, D., 110, 111, 122
Mulhall, S., 193, 195
Muller, G.E., 64
Munier, R., 141

Nagel, T., 173, 176, 185, 193, 195
Natanson, M., 100
Naville, P., 140
Nehamas, A., 181, 195
Nenon, T., 103, 104, 108, 113, 121, 122
Nietzsche, F., 148, 161, 169, 171, 178, 181, 182, 189, 191, 192, 193, 195
Nijinsky, V., 153
Nissl, F., 151, 153

Nussbaum, M., 187, 195
Nyström, M., 38, 42, 98, 100

Oatley, K., 187, 193, 195
Okrent, M., 112
Ott, H., 139, 144

Palmer, R.E., 25, 43, 99
Pascal, B., 139
Petersen, P., 151, 168
Petzet, H., 139, 144
Philipse, H., 115
Phillips, J., 138, 144
Pinel, P., 150, 168
Plato, 135, 140, 149
Plous, S., 193, 195
Pollio, H.R., 38, 43
Popper, K., 191, 195
Pribram, K.H., 16, 43
Protagoras, 164

Raabe, P.B., 195
Rank, O., 170, 172, 191, 192, 195
Ratcliffe, M., 193, 195
Regnery, H., 138
Rickman, H.P., 158, 168
Ricoeur, P., 25, 26, 43, 170, 204, 205
Rilke, R.M., 166
Robinson, E., 142, 157
Robinson, G.D., 25, 43
Rogers, C., 36, 43, 70, 176, 195
Rorty, A.O., 195
Rorty, R., 166, 168, 174, 180, 187, 193, 195
Rumpel, H., 144
Runes, D.D., 49, 68
Rybalka, M., 139, 140, 143

Safranski, R., 139, 140, 142, 144
Sartre, J.-P., 6, 7, 8, 9, 26, 27, 28, 43, 46, 68, 101, 115, 116, 122, 123, 124, 125, 126, 127, 128, 129, 130, 131, 132, 133, 134, 135, 136, 137, 138, 139, 140, 141, 142, 144, 149, 169, 171, 172, 174, 175, 176, 177, 179, 180, 181, 183, 184, 189, 190, 191, 192, 193, 195
Schilder, P., 153
Schmitt, W., 158, 168
Schneider, K., 153
Schrader, G., 145
Schrift, A., 138, 145
Schutz, A., 38, 43

Searle, J., 111
Seligman, M.E.P., 74, 88
Selz, O., 64, 65
Sexton, V.S., 35, 36, 43
Shouery, I., 145
Skinner, B.F., 16, 17, 43, 199, 205
Smith, B., 103, 111, 115, 122
Smith, D.W., 103, 111, 115, 122
Smith, J.A., 32, 44, 203, 205
Smith, J.E., 25, 43
Snygg, D., 3, 35, 41, 42, 44, 70
Solomon, R., 187, 195
Spiegelberg, H., 45, 68, 145
Spinelli, E., 38, 44
Stambaugh, J., 142
Straus, E., 142
Stumpf, C., 58
Szasz, T., 170, 189

Taylor, S., 182, 183, 195
Thompson, C.B., 38, 43
Tillich, P., 169, 170, 172, 174, 179, 180, 181, 191, 192, 195
Titchener, E.B., 4, 54, 56, 57, 58, 59, 60, 61, 63, 64, 65, 66, 67, 68
Towarnicki, F., 142

Valle, R.S., 38, 44
van Buren, J., 102, 105, 106, 107, 117, 119, 121, 122
van den Berg, J.H., 38, 44
van Deurzen-Smith, E., 8, 170, 173, 177, 181, 187, 188, 190, 191, 193, 195
van Kaam, A., 70, 170, 171, 172, 175, 195

Waelhens, A., 139
Wahl, J., 142, 145
Watson, J.B., 14, 16, 44, 46, 65, 68
Wertheimer, M., 30
Wertz, F., 37, 44
Wheway, J., 177, 178, 196
Wilber, K., 192, 196
Wilcox, R., 145
Wilson, S., 74, 87
Witasek, S., 58
Wittgenstein, L., 41, 154, 156, 166, 168, 170
Wolfe, B.E., 74, 87
Woodworth, R.S., 54, 68
Woolf, V., 178, 196
Wright, R.H., 74, 87

Wundt, W, 4, 12, 13, 14, 29, 34, 44, 46, 54, 55, 56, 58, 60, 61, 64, 65, 66, 67, 68, 151

Yalom, I.D., 8, 138, 145, 170, 172, 173, 175, 176, 177, 178, 179, 180, 183, 184, 185, 186, 187, 189, 191, 192, 193, 196
Yeats, W.B., 180, 196
Yerkes, R.M., 54, 68

Zemlich, M.J., 74, 87
Ziegler, D.J., 182, 194
Zuckerman, M., 192, 196

SUBJECT INDEX

absurdity, 183–187
act psychology, 13, 57, 58, 63–66
action research , 31
adaptation, 61
adumbration, 48, 52, 53, 54, 63, 115
agency, 38
alienation, 8, 172, 173, 190
aloneness, *see* isolation
anxiety, dread, 9, 26, 36, 83, 134, 172, 173, 174, 175, 176, 179, 180, 183, 186, 187
apparent movement, 29
appearance, the apparent, the phenomenon, 20, 21, 25, 28, 34, 38, 41, 51, 62, 75, 79, 82, 93, 95, 115, 149, 198, 199, 200, 202, 203, 204
appearance as adumbration *see* adumbration
apperception, appresentation, 93, 94, 113
atomistic-reductive bias, 34
attribution theory, 31, 182
authenticity, 9, 36, 111, 118, 151, 172, 174, 176, 178, 179, 181–183, 185, 187, 189
awareness, manner of being aware, *see noesis*
awareness, object of awareness, *see noema*

bad faith, *see also* self deception, 7, 27, 127, 128, 172, 189
behaviorism, science of behavior, 2, 12, 14, 15, 16, 17, 31, 32, 36, 46, 61, 65, 131, 199
Being-in-the-world, *Dasein*, 8, 24, 25, 26, 107, 111, 112, 117, 118, 119, 130, 133, 134, 137, 148–167, 201, 202
bifurcation of public and private worlds, *see also* dualism, 17
body, corporality, embodiment, lived body, body subject, 27, 28, 38, 39, 40, 108, 112, 113, 115, 120, 160, 161, 173, 174, 175, 183, 187
bracketing, *see epochē*
bridling the world, *see also epochē*, 92, 94, 98, 99

care, concern, 26, 107, 109, 134, 162, 173
causality *see* determinism
cognitive over-emphasis, 6, 102, 106, 107, 108, 109
commonsense psychology, 31
comportment, 106, 108, 109, 110, 111, 113, 115, 163, 201
consciousness, *see also* experience intentionality, 3, 4, 13, 14, 15, 16, 17, 27, 36, 45, 47, 48–51, 53, 54, 55, 57, 58, 59, 60, 61, 62, 63, 66, 67, 69, 78, 84, 89, 92, 93, 95, 97, 99, 100, 108, 114, 115, 116, 119, 120, 121, 125, 190, 199, 200
consciousness as 'internal,' 7, 12, 15, 17, 47, 115, 157
consciousness as a research standpoint, 1, 2, 3, 4, 9, 11, 14, 16, 17, 19, 36, 45, 46, 48–51, 54, 60–67, 84
consciousness and world split, *see also* dualism, 110, 115, 118, 119, 200, 201
constancy hypothesis, 29, 30
content psychology, 57
culture, the social, 8, 25, 28, 38, 77, 97, 98, 105, 106, 112, 188

Dasein, *see* Being-in-the-word
Daseinsanalysis, 8, 148–167

death, 8, 9, 38, 129, 134, 172, 175, 178–181, 184, 185, 187
determinism, causality, 7, 36, 49, 50, 51, 53, 56, 154, 156, 157, 159, 162, 165, 166, 167, 171, 178, 182, 183, 197, 204
discourse, *see also* language; speech, 2, 41, 204
doxic thesis, 109, 110
dualism, anti-dualism *see also* consciousness and world split, 4, 7, 16, 17, 19, 23, 28, 49, 56, 61, 62, 118, 173

ego, empirical/mundane, *see also* self; identity, 6, 27, 34, 47, 63, 114–118, 202
ego, splitting of the ego, 7, 110, 114–118
ego, transcendental, 6, 27, 34, 98, 99, 102, 107, 114–118, 202
epochē, 3, 4, 5, 14, 18, 20, 21, 28, 30, 34, 37, 38, 41, 51, 53, 82–84, 91, 92, 94, 98, 99, 102, 114, 167, 199, 201, 202–204, 205
empirical, 'real' object, 50, 52, 53, 56, 65, 83, 115
empiricism, 45, 48–51, 53, 54, 55, 56, 58, 61, 62, 63, 66
environing world, surround world, *see umwelt*
essence, *eidos*, 18, 21, 22, 23, 24, 25, 36, 37, 38, 47, 48, 50, 52, 53, 64, 65, 66, 75, 78, 106, 115, 131, 132, 133, 134, 135, 156
essence, *eidos* of disciplines, 18, 21, 22, 24, 25, 198
essentialism, 24, 27, 106, 178, 202
être-en-soi (being-in-itself), 7, 125, 126
être-pour-soi (being-for-itself), 7, 125, 128
evolution, 60, 61, 154, 171, 172
existential analysis, *see also* existential psychoanalysis; see also *Daseinsanalysis*, 8, 170, 178
existential dimensions, 8, 172–175, 188, 190
existential phenomenology, 3, 7, 17, 21, 22, 24, 25, 26, 27, 28, 41, 103, 188, 197
existential psychoanalysis, 7, 28, 123–137, 171
existentialism, existentialist, 8, 21, 22, 24, 28, 33, 36, 102, 106, 115, 120, 121, 130, 132, 133, 134, 135, 169–191, 197
Existenz, 129, 130, 132, 133, 134, 135, 137
experience, 4, 11, 28, 29, 50, 52, 54, 55, 56, 60, 71, 77, 78, 80, 83, 157, 175, 199, 201, 203, 205
experience as part of the objective world, 1, 2, 4, 12, 55, 56, 199
experience as basis of concepts, 18, 28, 29, 198
experience, lived, *see also* lifeworld, 5, 6, 9, 16, 28, 29, 41, 74, 75, 77, 79, 80, 81, 82, 83, 86, 98, 106, 107, 114, 174, 205

first person perspective, 15, 33, 34, 36, 41, 177, 198
flesh of the world, 95
force field, 31
free imaginative variation, 21, 70, 73, 85, 86
freedom, 7, 8, 9, 36, 38, 120, 121, 125, 128, 131, 136, 172, 174, 175, 177, 178–181, 185, 187, 188
fulfilling acts, fulfillment, 84, 85, 110, 111, 116
functionalism, functionalist psychology, 4, 57, 58, 60–66

genetic bias, 34
general description, general structure, 37, 73, 75
Gestalt psychology, 3, 13, 14, 28, 29, 30, 31, 32, 33, 34, 41, 65, 67, 197
guilt, 9, 159, 160, 175, 178–181, 184, 185, 187

hermeneutic phenomenology, 3, 24, 25, 26, 156
hermeneutics, 26, 92, 99, 105, 106, 156, 204
hermeneutics of suspicion, 26, 204
hermeneutics of meaning recollection, 26, 204
history, historicity, 8, 25, 28, 38, 77, 97, 105, 106, 108, 133, 135, 173, 187, 188, 198
human condition, 8, 131, 171, 174, 176, 187
human reality, 124, 125, 126, 127, 129, 130, 131, 132, 133, 134, 135, 163, 175
human relationships, social relatedness, 9, 15, 27, 70, 74, 75,76, 77, 82, 108, 163, 164, 173, 175–178, 179, 183, 189, 190
human science, *see also* psychology as human science, 1, 74, 89, 90, 91, 96, 99, 105, 205
humanism, 8, 32, 130, 131, 132, 133, 150, 161
humanistic psychology, 3, 7, 35, 36, 176, 178, 197

identifying acts, 84, 85
identity, *see also* self; ego, empirical/mundane, 39, 128, 135, 175, 184
idiography, 3, 15, 32, 33, 37, 41
illusion of the double world, *see also* dualism, 17
immanent object of mental acts, 48–51, 52, 53
integrity, 181–183
intellectualism, over-intellectualisation, 6, 110, 113–114
intentionality, 4, 5, 6, 7, 13, 15, 19, 23, 28, 41, 45, 47, 48, 49, 52, 53, 54, 56, 58, 59, 60, 61, 63, 66, 67, 78, 79, 84, 93–94, 95, 99, 102, 103, 105, 107, 109, 111, 112, 113, 114, 119, 120, 167, 171, 172, 199, 202, 203
intentional act, 7, 48, 83, 107, 111
intentional agent, 7, 119

intentionality, neglect of its being, 7, 110, 119, 120
interpretation, 17, 19, 25, 26, 29, 59, 82, 85, 86, 106, 156, 158, 159, 160, 203, 204
intersubjectivity, 5, 15, 76–78, 81, 82, 103, 170, 175, 176, 177
interview, interviewing, 34, 35, 37, 38, 71, 75, 76, 77, 80, 81, 82, 83, 198, 203
intrasubjectivity, 76, 77, 82
introspection, 13, 14, 15, 17, 31, 55, 56, 59, 60, 63, 64, 65, 66
intuition, 4, 5, 20, 21, 22, 50, 51, 52, 54, 65, 66, 67, 79, 84, 102, 103, 105, 106, 112
intuition of essences, 21, 23, 102, 106, 109
irreal, 50, 52, 53, 60, 63, 64, 66, 67
isolation, loneliness, 8, 160, 172, 174, 180, 187

Journal of Phenomenological Psychology, 3

language, speech, 25, 41, 152, 154, 155, 156, 161, 164, 166, 177, 204
lifeworld, world, 3, 5, 6, 26, 28, 36, 37–42, 61, 67, 74–76, 90, 91, 92, 93, 94, 95, 96, 97, 98, 103, 105, 106, 109, 110, 112, 118, 157, 162, 202
lifeworld 'fragments,' 38–41

manner of being aware, *see noesis*
meaning, 9, 15, 26, 28, 36, 41, 59, 63, 73, 75, 77, 78, 80, 86, 90, 91, 94, 95, 104, 109, 127, 156, 175, 177, 178, 198
meaning (existential), 183–187, 191
meaning unit, 37, 71, 72, 86
meaninglessness (existential), 8, 90, 172, 175, 180, 185
mood, 38, 124, 134, 157

natural attitude, 5, 6, 20, 50, 91, 93, 94, 95, 98, 108, 113, 118
naturalism, 48–51, 52, 54, 58, 61, 66, 90, 114, 119
natural science, 2, 4, 6, 8, 16, 20, 28, 46, 48–51, 54, 55, 56, 57, 59, 66, 70, 75, 87, 90, 91, 109, 110, 118, 120, 148, 151, 155, 156, 158, 162, 164, 165, 167, 197, 199, 205
Nazism, 148, 149, 157
noema (object of awareness), 19, 23, 24, 27, 28, 45, 47, 52, 54, 65, 66, 78, 107, 115, 199, 200, 201
noesis (manner of being aware), 19, 23, 24, 27, 28, 45, 47, 52, 64, 65, 66, 79, 107, 109, 115, 200, 201

objectification, 78, 131
objectifying acts, 113, 120
objectivity, 14, 15, 17, 59, 76, 77, 78, 89, 162, 183, 185, 190, 199
object of awareness, *see noema*
ontology, ontological, 8, 22, 27, 96, 102, 104, 105, 106, 109, 113, 118, 119, 120, 125, 126, 127, 128, 129, 131, 132, 134, 137, 148, 159, 162, 164, 166, 167, 175
original choice, fundamental project, 27, 28, 125, 126, 128, 129, 136, 137

perceptual approach, 15, 35
phenomenological psychological method, 5, 37–41, 45, 69–87, 188
phenomenological psychology, phenomenologically-based psychology;, 2, 4, 5, 15, 16, 17, 18, 25, 28, 36, 37–41, 47, 64, 69–87, 124, 198, 199, 201, 202, 204, 205
phenomenology characterised, 3, 17, 18, 21, 22, 25, 45, 53, 54, 63, 64, 66, 105, 106, 108, 149, 158, 169
phenomenology as a foundational discipline, 18, 91, 198
phenomenology, transcendental, 89–100
phenomenon, *see* appearance, the apparent
positivism, 16, 90, 151, 159, 197, 205
post-modernism, 9, 24
powerlessness, 38
practical action, *see* comportment; theoreticism
pre-predicative awareness, 103
pre-reflective, 5, 6, 17, 79–82, 108, 112, 202
pre-understanding, 94
presence, personal, 38
presence as conscious, 50, 51, 52, 53, 82, 91, 120
presentifying, making present, *see* intuition
project, 27, 28, 36, 39, 40, 125, 128, 129, 135, 136, 137
psychoanalysis, 7, 8, 25, 32, 46, 123–130, 131, 134, 135, 136, 137, 148, 153, 155–167, 170, 171, 172, 181, 189
psychological phenomenology, 31–34, 70
psychologically sensitive expressions, 72, 73, 77, 85
psychology, 1, 2, 11, 12, 51, 54–67, 62, 63, 151, 160, 169, 174, 177, 182, 199, 204
psychology as human science, 5, 11, 74, 90, 198, 199, 205
psychology as natural science, 4, 14, 16, 28, 46, 48–51, 54, 55, 57, 60, 63, 66, 70, 74, 197, 199

psychotherapy, therapy, 7, 8, 36, 74, 83, 124, 126, 128, 135, 136, 148–167, 169, 175, 177, 178, 181, 186, 189, 190

qualitative research, 1, 14, 15, 16, 33, 34, 35, 41, 87

real object, *see* empirical, 'real' object
reduction, 4, 6, 51, 53, 72, 77, 83, 89, 91, 96, 97, 99, 102, 107, 117, 118, 121, 199, 200, 201, 202, 203, 204, 205
reduction, psychological phenomenological, 83, 84
reflexivity, reflection, 63, 66, 94, 95, 96, 99, 102, 112, 115
reflexivity in empirical research, 96, 203
responsibility, 8

science *see* natural sciences
scientism, 1, 169, 190
schizophrenia, dementia praecox, 152, 153, 157, 163, 167
self, selfhood, *see also* identity; ego, empirical/mundane 13, 14, 27, 33, 34, 38, 39, 41, 161, 170, 172, 173, 174, 178, 180, 182, 183, 187, 202
self-awareness, self reflection, 94–96, 99, 100, 102, 172, 173, 174, 175, 179, 180, 181, 183, 186, 190, 202
self-deception, *see also* bad faith, 35, 181
sense of the whole, 71
signifying acts, 84, 85
situated structure, 37, 75
sociality; other people, *see also* human relationships, social relatedness, 38, 39, 175–178, 183, 184
solipsism, 92, 102, 107, 133, 184
spatiality, 38, 40, 154
stimulus error, 59
stimulus-receptor bias, 34
stream of consciousness, 13, 41, 47, 48, 49, 62, 115, 116
structural psychology, 55–60, 67
subjectivity, 2, 15, 16, 17, 31, 33, 34, 55, 63, 77, 78, 116, 130, 132, 171, 174, 185, 188, 190, 201, 202, 203, 205

teleology, 58
temporality, 38, 40, 134, 154, 162, 164, 174
theoreticism, over-theoretization, 6, 110–113
transcendent object of mental acts, 48–51, 52
transcendental subjectivity, 5

ultimate concerns, 8, 172–175, 188, 190
umwelt, 103, 104, 106, 108, 110, 113, 201
uncanniness, not-at-homeness, 8, 173, 174, 175, 176, 178, 181
unconscious, 7, 46, 78, 125, 127, 154, 165, 171, 180, 190
understanding, 2, 5, 99, 110, 112, 125, 202
use-value, use-objects, 103, 104, 109, 118, 162

value(s), 4, 7, 32, 36, 102, 103, 109, 110, 113, 114, 131, 178, 184, 185, 188, 190, 205
validity, legitimacy, 20, 21, 22, 74, 78, 96, 182
voice, 38
vorhanden, 103

willing, will, 8, 179, 181
world as 'external,' 7, 12, 15, 20, 190, 205
world as lived, *see* lifeworld

zuhanden, 103, 114

MIX
Papier aus verantwortungsvollen Quellen
Paper from responsible sources
FSC® C105338
www.fsc.org

If you have any concerns about our products,
you can contact us on
ProductSafety@springernature.com

In case Publisher is established outside the EU,
the EU authorized representative is:
Springer Nature Customer Service Center GmbH
Europaplatz 3, 69115 Heidelberg, Germany

Printed by Libri Plureos GmbH
in Hamburg, Germany